Who done it?

101 CASE STUDIES IN CONSTRUCTION MANAGEMENT

2015 MANUSCRIPT EDITION
LEN HOLM

COPYRIGHT 2001

INTRODUCTION

This manuscript began several years ago with a collection of just a few short case studies. There was a desire within the Department of Construction Management at the University of Washington to strengthen the students' written and verbal communication skills. These case studies were first used to improve the students' presentation skills within a Project Management course and also to provide practice for the subsequent quarter's Capstone Course. At that time the Department was just becoming active in regional student competitions. The activities of research, working in a team environment, preparing both a written and verbal response, and competing amongst fellow students helped the Department raise their competition teams to national recognition.

A few case studies were added each year. This manuscript now includes 101 case studies with sub sections for several cases and numerous questions for each one. The source of the cases are mostly from projects which I have personally been involved with; from 40 years in the construction industry as a carpenter, project engineer, project manager, owner's representative, construction consultant, and expert legal witness. A few of the topics have been donated from other friends in the industry. Several industry professionals and adjunct professors at the University of Washington have been using and supplementing to this body of work for two decades. Mike Matter, Sara Angus, Chris Angus, and James Shaiman have all used this material in their classrooms. I would like to thank them for their valuable contributions.

Abbreviations are used throughout the construction industry. Many of the abbreviations utilized within this work have been referenced in Appendix A.

As indicated, the initial purpose of compilation of this material in one manuscript was for the benefit of all students and the faculty at the University of Washington. Now the material is utilized within several construction management university programs and also for in-house contractor training. The uses are unlimited as discussed on the next page. I hope you all enjoy the stories.

Len Holm

USES FOR THIS MANUSCRIPT

There are many potential uses for this material; below are listed just a few. With some individual creativity from an instructor or construction firm facilitator, the possibilities are unlimited.

Writing Course: Case studies could either be assigned to, or chosen by, students for development into an essay type of response. The students' work could be evaluated both on the accuracy of their responses as well as their writing skills.

Competitions: Student teams can be assembled and set up to compete. Many of the cases are written in the form of "who is right...owner or contractor?"

Verbal Presentation Skills: Similar to the competition use, either teams or individuals can prepare verbal responses, including written or electronic media backup, to present their position on a particular issue.

Tests: With 101 cases, and over 80 additional subsections, and maybe 4 to 8 questions per case study and subsection, there are several hundred "short" questions which can be used on tests or for written homework assignments. An entire case study could be used for a "longer" written essay type of test question.

Project Management Class: This material has been successfully used to interact with the students after they have read the text and listened to a lecture on any one subject. For example the students are to read about claims before class. They hear a short lecture, which re-enforces the text. Maybe after a break, either again in teams or individually, they now present responses and debate a case or two that applies to that specific topic, in this case claims. Note lines are included at the end of each case for notes and responses.

Applied Case Study: Some university construction management programs utilize just one complete construction project as their applied case study. The students then use that one project in all of their course work. The estimates and durations noted in many of the 101 case studies included here could be modified to correspond with a program's applied case study.

Capstone Course: The student or the instructor could apply many of these cases to a specific project the student is working on in a capstone course. Analyzing these problems and either informally during a class discussion, or formally through a presentation, will help the students better prepare for industry evaluations on their projects.

Research: Upper division or masters students as well as industry professionals could take one or more of the case studies and research AIA contracts, laws, similar projects, legal cases, and interview contractors to prepare more lengthy and substantiated responses. In this case responses would include tables, graphs, statistics, and referenced materials to support their research and conclusions.

Unlimited Variations: An instructor or corporate facilitator could easily add a single line to any of these cases to customize the material for their course or company or desired outcome. Several questions are included for each case study. Many others are possible which could allow this material to expand exponentially.

Professional Development: Contractors can apply many of these examples to their own projects. "Lessons Learned" exercises presented during in-house corporate training sessions or professional seminars would be very beneficial.

TABLE OF CONTENTS

The 101 case studies are generally organized according to their primary topic or 'Section' as indicated below. The order of the 14 sections follows the textbook *Management of Construction Projects: A Constructor's Perspective* published by Prentice Hall and authored by John E. Schaufelberger and Len Holm.

Section	General Topics	Page
1.	Organizations: Including Project Organizations, Company Organizations and Personnel Issues	6
2.	Procurement	16
3.	Contracts: Including Insurance and Bond Issues	35
4.	Estimates	53
5.	Schedules	58
6.	Subcontractors: Including Subcontracts, Purchase Orders, and Supplier	63
7.	Startup: Including Pre-Construction, Mobilization, and Value Engineering	77
8.	Communications: Including Documents, Documentation, RFIs, Submittals, Written and Verbal Communication Skills	83
9.	Pay Requests: Including Liens and Lien Releases	89
10.	Cost Control	94
11.	Quality Control	98
12.	Change Order Proposals: Including Contract Change Orders	106
13.	Claims: Including Dispute Resolution Techniques	113
14.	Advanced Topics: Including Close-Out and Safety	121
Appendix 1:	Abbreviations	126
Appendix 2:	Case Study Matrix	129

There are several case studies which fit within each of the above primary sections. Each of the 14 sections begins with its own detailed table of contents. Many cases relate to other sections as well; most of the examples cross many topics. Appendix 2 at the back of this manuscript lists all 101 of the cases in a table format and cross-references each individual case with the fourteen primary topics. See Appendix 2 for additional clarifications.

SECTION 1: ORGANIZATIONS

Including Project Organizations, Company Organizations, and Personnel Issues

Case	Title
1.	Pass-Through Contractor
2.	Owner's Contracts
3.	Construction Manager or General Contractor?
4.	Design-Build Joint Ventures
5.	Owner's Subcontractors
6.	Bait and Switch
7.	Multiple Contracts
8.	Ambitious Project Manager
9.	New Contractor
10.	Client Expertise
11.	Developer or General Contractor?
12.	Overworked Project Engineer
13.	Self-Performed Failure

Most of these case studies overlap with other primary topics and at least 10 other cases involve organizations. See Appendix 2 for a matrix connecting all 101 of the cases with all fourteen primary topics.

Case 1: PASS-THROUGH CONTRACTOR

A commercial construction firm which reports to be a design-build general contractor actually operates more as a construction manager (CM). They sell their services as being an agent for the owner. This firm will procure all design and construction services, originate the contractual agreements, charge a fee on all hard and soft costs, but have the owner and the second tier firms execute the contracts direct. This method of procurement is referred to as a "pass-through" contract. The CM does not sign nor initial any of the contract documents, pay requests, change orders, field questions, submittals, etc. The CM explains to owners the attractiveness of this contracting arrangement as a means of saving additional tax and insurance markups. The CM does not have problems convincing the subcontracting and supplying firms of the advantage of contracting direct with an owner as these firms are now one step closer to the owner and subsequently the bank. Several of this CM's projects have had problems with second tier contractors (such as foundation settlement, lack of sufficient air conditioning, window leakage, roof leakage, or dead landscaping). In each instance the CM has stepped back, not protecting the owner, and force the owner to resolve problems with subcontractors direct. On other projects, there have been problems with the owner not paying the contractors or designers and again the CM has stepped back, not protecting the subcontractors. The second tier firms have had to pursue resolution, through liens or other means of collection, with the owner direct. The CM receives a guaranteed lump sum fee for their services and jobsite general conditions are cost reimbursable. What is wrong with this system? Does it happen? How can it be improved and still utilize the services of a construction management or owner's representative firm or individual? How would a general contractor or CM "at-risk" system improve this condition, if at all? Develop three organization charts (pass-through, owner's rep, and CM at risk) and utilize them in answering these questions.

Case 2: OWNER'S CONTRACTS

2.1 On a very large, very complex remodeling and expansion project an owner has contracted with numerous members of the design team independent of the architect. The architect does not have any second tier design firms under their contract. The owner has also contracted with many different suppliers and specialty subcontractors direct and not through the general contractor. The owner is of the belief that they can save on multiple markups by doing business this way. The owner has although, written into every contract, that each firm is still responsible to "coordinate" and communicate direct with the other independent designers and contractors as if they were contracted under the standard vertical method rather than this horizontal method. What risks has the owner assumed? What risks are the designers and contractors assuming through this system? Draw up organization charts depicting this condition and the "standard" condition.

2.2 Assume the following costs. What would the cost increases have been to the owner if they had contracted in a normal fashion? Does this also provide "contracting opportunities" to the contractors or designers? Explain why or why not.

- $50 mil total construction (or utilize an available construction estimate)
- $30 mil total construction cost of sub-consultant designed work
- 4% total architectural fee on the architectural portion of the work and they would receive an additional 10% markup on sub-consultants' design fee if run through the architect's books
- The sub-consultants would receive 10% design fee of the construction cost of their portion of the design, either billed direct to the client or through the architect
- 5% general contractor fee on costs run through their books
- The values for four sample subcontracting firms working direct for the owner are $550k, $950k, $5 mil, and $12 mil. Assume the subcontractor fees are included in these figures. The other subs traditionally work through the GC

Case 3: CONSTRUCTION MANAGER OR GENERAL CONTRACTOR?

A client is ready to negotiate a contract with a construction firm for a $30 million shelled office building project. The design-development (DD) documents are complete. The building permit has been applied for and is scheduled to be issued in two months. The architect has requested the owner now bring on a contractor to assist with the balance of pre-construction services, estimating, scheduling, constructability analysis, material selections, and value engineering (VE) during the construction document (CD) development phase. The client and the architect have received written proposals and conducted interviews and have narrowed the short list down to two firms who have a completely different approach to contracting. Both appear to be equally qualified with respect to experience, references, availability, etc. Both firms have worked with the architect and the owner successfully on previous projects. Both firms are quoting a competitive 4% fee on top of the cost of the work. All other conditions are equal. The only difference between the two firms is that one is a pure construction manager (CM) and will subcontract 100% of the project except jobsite administration. The other is a typical general contractor (GC). The GC is only interested in building the project if they are allowed to perform the work which they customarily self-perform, such as concrete, carpentry, reinforcement steel, structural steel, and miscellaneous specialty installation, which will account for 30% of the cost of the work on this shell.

3.1 Take the position of the CM. Why is to the owner's and the architect's advantage to employ your firm during the pre-construction phase? What are the advantages of using a CM during the construction phase? Discuss all project control issues including quality, safety, schedule, and cost. Is an 'at-risk' CM truly an owner's representative? Why is it better for you that your firm is selected now and on-board when the tenant improvement projects become available? Does the GC "hide" costs? Sell your position and be creative. Use the tools you have learned from this manuscript, your classes, professional experience, and outside research to convince the owner that the CM procurement approach is more advantageous than the typical general contractor.

3.2   Take the position of the GC. Cover the same issues as outlined in the CM position above. Does the GC have more control over cost, schedule, quality, and safety? Which firm can build it better, faster, and safer? Who best looks out for the client's interests?

___

Case 4:   DESIGN-BUILD JOINT VENTURES

A client negotiated a design-build contract with a joint venture (JV) general contractor and architect for the design and construction of a $100 million chemical manufacturing facility. The project was constructed within a reasonable time frame and for a fair price. The quality of the work was acceptable and there were not any time-loss safety incidents. The architect's and the contractor's joint venture was dissolved immediately after substantial completion was achieved and they had received their retention. In only six months after completion the owner had achieved 90% of production output potential and sales were higher than anticipated. About that time, several employees began feeling ill and several citizens living and working outside the mill filed a claim due to unacceptable odors coming from the plant. The government shut the plant down. It was discovered that there was a major design error in the mechanical exhaust system and it would cost the owner $10 million to re-design and repair and maybe ten times this in lost revenue. There would eventually be personnel claims as well due to illness. The owner took the designer to arbitration and won. The owner separately took the contractor to arbitration and won on the basis that the general "should have known" that the design was in error. Should the owner be allowed to pursue the two parties separately? Was the general at fault? Shouldn't the owner have had some responsibility in this? The city approved the plans - what sort of liability do they have? What types of insurance will come into play here? Does this mean the design-build procurement approach is flawed? What sorts of checks and balances are required of the design for a design-build project?

Case 5: OWNER'S SUBCONTRACTORS

The prime general contractor (GC) on this site is a union firm (some union affiliations and predominantly union subcontractors). The contract allows the owner to employ subcontractors direct, which they may do for a variety of reasons. If the owner's contractors are open shop, but not labor trades which the union GC or their subcontractors are affiliated with (such as painting), is there a problem? What if the labor trades are similar, such as the owner hiring an open shop electrician and the GC contracting with a union electrician? How does a "client friendly" GC manage these scenarios?

Case 6: BAIT AND SWITCH

This contractor conducted considerable research before submitting their negotiated proposal to the owner and architect teams. The GC proposed management and supervisory individuals that were both familiar to the reviewers and had relevant project and location experience. The market was very busy. After award and contract execution the GC systematically changed out all four members of the team who were proposed and participated during pre-construction. Each change was explained and apologized for. Can they do this contractually? Is this ethical? Is this "bait and switch" customary on negotiated projects? Who loses when new team members are brought on board? When might it be to either the contractor's or the owner's advantage to bring on new participants?

Case 7: MULTIPLE CONTRACTS

Some owners will contract with multiple general contractors on one site. This scenario is sometimes referred to as "multiple primes". Some owners will employ specialty subcontractors direct. What sort of risks and coordination issues is an owner assuming by not placing all of the work on one site under one GC? Does this save or cost the owner money? Do these sorts of multiple contracts also place risks on the contractors or designers? Does this also provide for "contracting opportunities" for the contractors?

Case 8: AMBITIOUS PROJECT MANAGER

This relatively new yet very ambitious general construction (GC) Project Manager (PM) was given an assignment for a repeat industrial client. The project was bid lump sum at $50 million by a staff estimator. There was a $1,000,000 bid error. The Officer-in-Charge (OIC) directed the PM to get every change order possible. The PM was not to worry about the possibility of future work with the client. During the one year of construction, the PM lost his project engineer and was not allowed to replace him with any experienced help. He had to hire from outside. He also did not receive any home office supervisory help. The OIC would meet him for lunch once a month at a remote site. Why did the OIC distance himself? The project was ultimately brought in with a clear fee of $600,000, but relations had been damaged; the GC did not get to work with the client again. Later in the PM's career with other contractors, he was not welcomed back on the client's site. The PM did what he was asked to do, didn't he? He was successful, wasn't he? Did the PM make any errors?

Case 9: NEW CONTRACTOR

A successful and experienced project manager has just left a very large construction firm and decided to become her own general contractor. How does she get started? What type of consultants does she employ? How does she find office and field help? What recommendations would you make so that she does not end up just waiting for the phone to ring? What is a 'no compete agreement'? Should she target clients in just one or several industries? What sort of banking, accounting, insurance, surety, and legal resources should she have in place? What would the start-up costs be?

Case 10: CLIENT EXPERTISE

10.1 During the course of construction of a 500-unit apartment complex, the developer has hired the project manager/estimator (PM1) away from the project's general contractor and also the architectural project manager (Arch1) away from the prime designer. Both of these people were involved in the project during pre-construction. Why would the developer do this? What complications could arise from this for all parties? Is this ethical?

10.2 You were previously the GC's PE and are now thrown into the position of the general contractor's new project manager (PM2). There is also a new architect (Arch2) who was previously a draftsman working for the old architect. He is a yes man and knows the way to get more work from the developer is to take their side on all issues. Your predecessor is also pushing you around from the client's side of the table, claiming everything was 'included' in the bid. How do you deal with all of these new "clients" who obviously know more about the project than do you?

Case 11: DEVELOPER OR GENERAL CONTRACTOR?

This case involves an employee owned general construction firm. The three major stockholders (your bosses) are also developers in commercial speculative offices. They of course always hire their own construction firm to construct their facilities, but keep all of the dollars separate. You have just been promoted to the position of project manager for the construction firm. You have been with them for three years as a project engineer. It is considered a compliment that they would ask you to manage their own development projects. The first project you are managing is a $5 million two story office building, which was estimated by a staff estimator. It appears that he has forgotten the extensive $100,000 millwork package that the developers would like in the lobby. Because the "client" is very tight with the development costs, they usually employ an architectural firm who prepares less than complete documents, relying on the general contractor to "fill in the blanks." The millwork is not drawn or detailed and is only vaguely implied in the outline specification book. You bring this situation to the developers who indicate that it is the contractor's problem and this millwork will be required. This will eat up half of your projected fee. Is this fair to you? Is it fair to the construction firm and the minority stockholders? How do you deal with this situation? Assuming that this process of moving expenses to the contractor's side of the ledger continues, do you consult with some of the other employees?

Case 12: OVERWORKED PROJECT ENGINEER

You are a new project engineer (PE) with a general construction firm. You are assigned to work out on a project jobsite with a very experienced superintendent who does not like to deal with paperwork. The tasks assigned to you by the project manager include items such as:

- Expediting materials
- Handling the field question and submittal logs
- Approving invoices
- Researching backup for change orders
- Taking and distributing meeting notes

The superintendent is delegating many of his tasks to you as well. He requires you to:

- Complete the daily diaries
- Write all of the field questions generated by the foremen or subcontractors
- Review all submittals and shop drawings
- Answer the phones
- Receive and sign for all material deliveries
- Chair the weekly toolbox safety meetings
- Update his and all of the foremen's drawing sticks with revised drawings and sketches
- Handle all short form purchase orders for materials

In addition, you are often in the pickup truck running errands. You didn't mind the six to seven day work weeks, or the ten to twelve hour days in the beginning, but your home life is certainly being impacted. You are learning a lot and this is why you wanted to get into the construction industry. Are you qualified to perform the tasks assigned by the superintendent? Should he be doing some of this himself? You bring this delegation issue to your project manager's attention but she indicates that you should just get along with the superintendent and not to rock the boat. The superintendent is tight with the president of the company and your PM warns, "the squeaky wheel could get replaced". It seems the only thing she does is write change orders, prepare the monthly pay requests, and the monthly forecast. Do you quit? What sort of risks is the company taking by placing all of this responsibility on your shoulders? What sort of risks are you taking? Do you still want to be a GC PE after graduation?

Case 13: SELF-PERFORMED FAILURE

A developer has negotiated with a general contractor (GC) to construct a mixed use (MXD), six-story, retail, office, and parking facility. The structure is a mixture of cast in place concrete, post tension concrete, and wood framing. The contractor is new to this location and has been subcontracting all of the work out and not performing any self-performed work. The contractor is criticized because of this by the local general contractors and accused of being a "suitcase" contractor. The GC would like to self-perform some work, but has not yet established a reliable local work force. The developer has hired an estimating consultant who reviews the contractor's estimate and finds that they received competitive bids in all normally self-performed areas and the subcontractor prices are consistently lower than are the contractor's direct estimates. Five months of construction pass and there is a failure with the post tension cables. It turns out that the contractor has self-performed all of the concrete work instead of subcontracting it out as was the plan. The contract is silent about requirements to use direct or subcontracted labor. It will cost over $300,000 to repair the damage. The contractor submits the repair cost to the developer's insurance carrier as the developer was carrying the builder's risk insurance. Is this an insurance claim? What is the best way for a contractor to become established in a new market and develop a work force? Does the developer have a claim to make against the contractor because of the estimate review and the "agreement" to subcontract the concrete work? If there are other savings in the GMP can they be used to cover these repairs?

Who Done It: 101 Case Studies in Construction Management

SECTION 2: PROCUREMENT

Case	Title
14.	Design-Build Mechanical and Electrical
15.	Marketing
16.	Bid or Negotiate?
17.	GC/CM Accounting
18.	Bid at 50%
19.	Union or Open Shop?
20.	Executive Home
21.	Successful Schools?
22.	Irrigation Union?
23.	Public Set-Asides
24.	Public Alternatives
25.	Labor Agreement
26.	Negotiated Success

Most of these case studies overlap with other primary topics and at least 11 other cases involve procurement. See Appendix 2 for a matrix connecting all 101 of the cases with all fourteen primary topics.

Case 14: DESIGN-BUILD MECHANICAL AND ELECTRICAL

14.1 This new athletic club involves very complex mechanical, electrical, and plumbing (MEP) systems, totaling over 40% of the entire construction budget. The owner has had a bad prior experience with a fully designed-bid-built mechanical project. Under the recommendation of the architect and estimating consultant, they decide to employ design-build (D-B) mechanical and electrical subcontractors direct. Do you agree utilization of D-B MEP subcontractors helps control costs? What role could a third-party MEP designer play with respect to 'criteria documents'? How much are design fees for MEP subcontractors? Who should carry these design contracts, the owner, architect, or the general contractor? Who carries the errors and omissions (E & O) insurance for this work? Who is named as additional insured? What risks do all of the parties assume?

14.2 Assume in this case, the client carries the design portion of the MEP contracts and a negotiated general contractor will ultimately carry the construction portion of the contract when a guaranteed maximum cost (GMC) contract is developed. The design agreements are very loose and the subcontractors' deliverables and detailed schedules of values (SOV) are not defined. The design team reports that the documents are complete and the general contractor submits their GMC to the owner. It is significantly over budget, especially in the areas of mechanical and electrical. The subcontractors refuse to negotiate their construction estimate and the owner announces they will bid this portion of the work out. Is this ethical? Is the owner guaranteed a lower construction estimate if this portion of the work is put on the street? Will the original designer-subcontractors submit competitive lump sum bids on their own drawings? If so, can they get change orders for discrepant documentation? What are the risks for all parties?

14.3   The design contracts are silent with respect to ownership of the design documents. The subcontractor-designers claim they own the documents. Because the owner has paid 90% of the design fee, with the final 10% invoice in process, the owner also claims it owns the documents. Who is correct? What would standard contract verbiage dictate?

14.4   A third party engineering firm and two subcontractor competitors review the drawings and agree that they are maybe only 50% complete and are not biddable. All of the details and equipment schedules are missing. The specifications are in outline form. The owner requests that the original contracted subcontractors finish the drawings. The MEP firms indicate that the owner has received what they paid for and that it is customary for the final details to be completed through the shop drawing and request for information (RFI) process. Are they right? If the documents are bid at this level will the industry "fill in the blanks?" If they are bid, who will take ownership of the design? How is this issue resolved? If the client does not pay the remainder of the 10% due on the contracts, can the design-build subcontractors lien the property even though they have not provided any material improvements?

Case 15:   MARKETING

General contractors must actively go out and pursue work or they will be limited to competitively bidding the lump sum public works market. The advantages of marketing are a) that they can negotiate a project with no competition, or b) can be placed on a short list for a negotiated proposal, or c) can be placed on a short list of comparable contractors for lump sum bidding. Minimizing the competition increases the contractors' opportunity for success. How can a firm which is new to a geographical area market themselves? How can a local firm market themselves into a new industry, such as a residential firm into retail construction? If a contractor is a small firm, how does the CEO make time available to market the firm? In the case of a larger contractor, what are the advantages of a staff "salesperson?" What are the disadvantages? When does marketing cross the line and become unethical and harmful to a firm's reputation?

___

Case 16: BID OR NEGOTIATE?

Two contractors are pursuing the same client for a dental clinic. The drawings are now 50% complete. The dentists' organization knows and trusts both contractors. The first contractor likes the lump sum market and is encouraging the client to complete the drawings and competitively bid the project. The second contractor is recommending that the owner allow a select group of contractors to propose on the drawings as they exist now and save additional architectural fees. 16.1) Argue the first contractor's case. What are the advantages to the owner to bid the project? 16.2) Argue the second contractor's case. What are the advantages to the owner to negotiate the project? Use statistics and materials obtained from your courses, text, and outside research to support your position.

___

Case 17: GC/CM ACCOUNTING

This public works project is being built under the GC/CM procurement method. This is also known by some as CM/GC. The GC/CM negotiates an early Maximum Allowable Construction Cost (MACC) with the client. The construction manager (CM) was also awarded the concrete and steel packages (work they normally self-perform as a GC) under separate competitive lump sum bids. The contractor is now a subcontractor to themselves.

17.1 How can the client be sure that costs incurred under the lump sum portions of the project by the CM's Subcontractor (Sub) team are not being charged to the CM's MACC? How are the CM and subcontractor kept at arms-length with respect to cost and contractual issues? Does the CM evaluate the Sub's change orders and pay requests critically? Does the CM help the subcontractor with site supervision and equipment usage more than they may with other subcontractors? Is this procurement system fair to the taxpayers? Are the CM's accounts auditable? Are the CM's accounts auditable for their own lump sum subcontracted portion of the work?

17.2 One reason many public agencies are choosing the negotiated GC/CM MACC approach over traditional competitive bidding with general contractors is to reduce claims and lawsuits. Can these results be validated? GC/CMs offer constructability reviews, value engineering, and early procurement of long-lead materials through this delivery method as well. List some of the other advantages. Can we determine the exact pay-back or return on investment for the client from these services? If you were assigned the task of researching and 'proving' these claims, what methodology would you choose?

Case 18:   BID AT 50%

A semi-conductor manufacturer has decided to build a new satellite facility. This project will eventually cost $200 million. It is their corporate policy to bid all work. Before a significant amount of design can be developed (soft costs spent) the Board must give approval on a maximum anticipated cost. This is a catch 22 situation as it is difficult to develop a maximum cost until significant design is available. The owner lacks in-house capability to develop their own estimates. At 50% design completion the owner solicits "bids" from four general contractors.

18.1 As a general contractor who has experience in this market, should you pursue bidding this project? Why or why not? If the designer is a reputable firm, with whom you have a good team relationship, does this change your decision? What alternate procurement methods could the owner consider?

18.2 The architect on this project recommends to the owner that they should use design-build subcontractors for the mechanical and electrical portions of the project early in the design process. Is this a good idea? Why would the architect suggest this? Should the owner follow this recommendation? If so, what sort of guidance or criteria would you recommend the owner follow in the selection process of these two important subcontractors? When should these firms be brought onto the team? What selection process should be followed: low fee, resume, construction estimate, design approach, others? As the general contractor, does this change your decision to pursue this project? Does the general contractor assume more or less risk with design-build subcontractors?

18.3. The semi-conductor client decides to save additional design fees and bid out the design-build subcontractors concurrent with the general contractors. Is this a good idea from the owner's perspective? Is this fair to either the subcontractors or the generals? How can the general contractor bids now be evaluated properly? Who carries the design portion of the contract and the associated errors and omissions (E&O) insurance?

18.4. As the general contractor pursing this project your firm decides to team up with a design-build mechanical and plumbing subcontractor during the bid process. You work together closely to assure that your scopes of work are consistent. The mechanical subcontractor's price is $80 million. On bid day you receive an unsolicited proposal from an outside mechanical firm which was teaming up with one of your general contractor competitors. This subcontractor is a very competent firm which has experience in this industry as well. Their proposal appears complete. Their price is $75 million. What do you do?

_____
_____
_____
_____
_____

18.5 List other methods this owner may have chosen to get valuable competent early estimates. Can any of these methods 'guarantee' a cost to the owner?

_____
_____
_____
_____
_____

Case 19: UNION OR OPEN SHOP?

You are presenting your company to a client who has a 50,000 square foot $10 million tilt-up concrete construction project to award. The project as-planned is now a "shell" with potential TI (tenant improvement) work to be negotiated at a later date. The client has utilized both union, and open shop (or merit shop) construction firms on prior projects. The client does not yet have a bias towards either choice of labor scenarios. The client has narrowed their contractor choice down to two relatively equal firms. Both contractors have extensive resumes in this type of work. Both contractors have successfully worked with this client previously. Which is the best system of labor, union or open shop? One team (19.1) is to assume the union position and argue their point; the other (19.2) is assigned the open shop position. Try to anticipate the points your competitor will make when preparing your own presentation. Convince the client that your firm has the best choice of labor. Use information presented in your text and classes as well as outside research to form the basis for your presentation. Cover subjects such as quality, schedule, flexibility, cost, training, and safety. Assume that the project is not being built in a labor-bias area, such as the northwest and northeast are predominantly union whereas the southeast is predominately open shop. Be creative. Only one team will be awarded the project. The other team will get nothing.

Case 20:   EXECUTIVE HOME

20.1. A Pharmacist and his wife have agreed to a purchase-sale agreement with a sole-proprietor contractor/developer for an executive, yet speculative, $5 million home. The contractor had not yet started construction when the agreement was signed. $200,000 is put into escrow for earnest money. All remaining funds are to be transferred upon closing, which will occur after certificate of occupancy (C of O). The buyers were provided with six pages of plans and an artist's rendering from which their agreement was based. There were not any separate specifications. The house is scheduled for a 12 month construction duration. What risks is the contractor taking? What risks are the buyers taking?

20.2. In a custom home scenario, the buyer employs the contractor to build their "dream." In a speculative situation, the buyer receives what the contractor intends to supply. In this case, the Pharmacist and his wife expect custom features and upgrades, considering the purchase price. The contractor is quick to point out at every turn that either a) the drawings said plastic laminate, not marble, or b) the 40 cents per square foot allowance for paint over drywall does not cover cherry wood wall paneling. The contractor requests change orders for all of the upgrades and the buyers agree. The purchase-sale agreement requires all change orders to be paid up front, in cash.

Is this a standard clause? Why does the contractor require this? Over the course of construction, almost $1 million will be added in change orders and paid up front by the buyers before closing. Now what risks do the two parties have?

20.3. During the construction process the residential construction market has improved dramatically. All of the new high-technology millionaires are driving up the sales prices of these executive homes. Approximately halfway through, the builder realizes he could have sold this speculative home for $7 million without the buyers' upgrades. With the upgrades (which have all been nice improvements, including a dock and wine cellar) the house may be worth over $8 million. The builder is regretting he put the home on the market so soon. The builder acts as his own project manager. He also has a full time on-site superintendent. Both of them do everything they can to make the process miserable for the buyers. Construction on this site noticeably slows down. They claim the busy market is taking away their subcontractors and quality craftsmen. They know the buyers have sold their previous home and are uncomfortable in temporary quarters. The buyers are really pressuring the contractor to speed up. The only thing they speed up is the original contract work that is certain to require rework due to contemplated upgrades by the buyers. This drives up the cost of the changes even further and adds to the buyers' frustration level. Why is the contractor causing these conflicts? What recourse does the buyer have? If the buyer walks, who keeps the earnest money? Who keeps the $1 million in change orders?

Case 21: SUCCESSFUL SCHOOLS?

21.1 This school district had a long record of working with repeat general contractors. Even though the state law required competitive bidding, the same general contractor would usually be low bidder. Quite often the second tier subcontractors were also the same from project to project. The architect and sub-consultants were also consistent on every project despite state requirements for competitive proposals. The contractor, architect, and the school district all thought quite highly of each other. After 10 projects together none of the parties had ever claimed each other. When their success was investigated by an outside consultant, he found that the client carried a fifteen percent contingency after the bids were received. Is this a reasonable amount considering that 100% plans and specifications were prepared? At the end of the project almost none of the contingency was left. What was happening here? Were the taxpayers being treated fairly? Were the competing contractors and designers being treated fairly? Isn't our ultimate goal a quality project without claims?

21.2 Many professionals in the construction industry feel designers are currently preparing less than complete documents today. The competition for projects has in many cases caused design fees to be lowered. This same school district asked their consultant to review their architect's fees and see if additional fees would result in better documents. As stated above, the district almost exclusively used the same design team. In response to the question, the architect indicated that they thought their fee was already fair. The state law determines that the architect's fee is approximately 7% and is not negotiable by either party. The law although does not discuss limits on change orders. What the consultant discovered was that the designer had actually been receiving a total fee of over 10% due to design services change orders upon project completion. Was this fair? Should this have resulted in more accurate documents? Is everyone better off by just letting this issue pass? How should design fees or designer selections be made on public projects? How are they made for private projects? Do you feel that lower design fees relate directly to poorer quality design? Can this be substantiated one way or another through research?

Case 22: IRRIGATION UNION?

You are the GC's PM working on a union construction project for a major industrial client who employs only union mechanics. There are very few union landscape and irrigation contractors. The irrigation subcontractor in this case is non-union. The plumbing subcontractor is union. All of the irrigation work to be performed on the project is outside of the building lines. The morning that the irrigation pipe is delivered to the job, the building union plumbers stop the delivery truck. The plumbers pull the driver out of the truck and physically assault him. The entire plastic irrigation pipe is pulled off the truck and busted up. Why did this happen? Who is at fault? How could it have been prevented? What do you do today to solve the problem?

Case 23: PUBLIC SET-ASIDES

23.1 The voters of this State have recently approved a new issue that abolishes preferential treatment, or 'set-asides' to minority groups. Explain what set-asides are. Explain the terms MBE/WBE/SBE/VBE/DBE. What were the pre-vote percentages of work that needed to be awarded to these groups on public projects such as a high school? How many dollars would these percentages represent on a $100 million project? How will this new legislation affect these types of firms? How will this new legislation affect the way these construction projects are bid, awarded, and constructed? What is the current status of this legislation in other states? Why are voters now over-turning these old laws?

What will be the outcome? Take either the pro set-aside position (23.2) and argue against this legislation or the open-markets position (23.3) and argue against state imposed minority percentages. Use information from your classes, text, and outside research.

Case 24:   PUBLIC ALTERNATIVES

24.1   Many public agencies such as the Federal Government, public universities, and cities and States are evaluating alternative procurement methods in an attempt to avoid the negative claim atmosphere that currently surrounds lump sum bidding. What are these alternative methods? Who favors each method (public, government, large contractors, small contractors, design community, minority contractors)? Do public bid jobs end up with more claims? Substantiate your position with facts from projects which utilized alternative procurement methods. What is going right with this movement? What is wrong? What can you recommend will solve these problems? Base all of your conclusions on information you have learned in this manuscript, your classes, and outside research. Be creative.

24.2   Argue in favor of the traditional open market lump sum bid approach. Is this truly a "bid 'em and sue 'em scenario?" Why is the public perception incorrect? Use facts and figures to support your position.

24.3   Argue in favor of short-listing qualified subcontractors and general contractors. How can short-lists be fairly prepared? Can this reduction of competition improve relations, quality, safety, cost, and schedule control? Does this restrict competition and drive up prices to clients? Which is most fair to the tax payers?

24.4   Prepare an argument in favor of the GC/CM alternative procurement method. Is this fair for all firms, including the taxpayer? Does the GC/CM have an incentive to under-staff the project? Will other traditional general contractors bid the structural subcontractor package to the GC/CM? Would the GC/CM treat other GCs fairly? If the GC/CM gets awarded the structural work as a lump sum subcontractor to themselves, how is the accounting of actual construction costs kept separate?

24.5   Prepare an argument in favor of the design-build (D-B) delivery method. Can this work? Is it working? Who provides the checks and balances to assure the successful firm is not "cheapening" the design to save on construction costs? Who approves the design documents? Who approves submittals? Who responds to RFI's?

24.6   Describe formal 'partnering'. Does mandatory partnering affect your position on any of the above delivery methods? Will it solve some of the problems? How much does partnering cost and which party pays for it? How much does it save?

Case 25:   LABOR AGREEMENT

A unionized public water district solicited bids for the construction of a $500 million water reservoir project. The request for quotations (RFQ) required all bidders to accept a "project labor agreement" which was included in the contract documents and negotiated between the water district and the local unions. This was a complex project and the water district was concerned about the qualifications from potential out of town open-shop firms. Bids were received from several "union" firms with the lowest at $490 million and one open shop firm at $480 million. The open-shop bidder was disqualified based upon the above requirements. The water district awarded the bid to the low union firm on the basis that they complied with the bid documents and their bid was within the project budget. The open-shop contractor took the water district to court on the basis that the "project labor agreement" violates state law which requires open and competitive bidding and the owner is required to hire the low bidder. Was the disqualification fair to the open-shop firm? Is this fair to the taxpayers? Will they be successful in over-turning this award? If the $480 mil firm wins the award, will the $490 mil firm have a basis for their own claim? How do government agencies generally address these situations in their bid documents? What are Davis-Bacon Wages or Prevailing Wages? What if this scenario was with a private utility company?

___

Case 26:   NEGOTIATED SUCCESS

The client, architect, and professional owner's representative team all agreed that bringing a negotiated general contractor (GC) on board early during the design process was the best procurement method for this project. Five contractors proposed on this private build-to-suit corporate office building. The proposals included:

- Individual team member resumes and similar corporate projects
- Safety records
- Financial references
- Current and anticipated volumes
- Milestone schedule and preliminary budget

26.1   The budgets were based upon early schematic documents. One contractor's budget was $18 million. Three were in the $22 to $25 million range. The fifth was high at $30 million. The contractor chosen had an average length schedule and a budget of $23 million. Why was the selection made for this construction team member? Why was the low estimator not chosen? Why do you think their budget was so low? Why was the high estimator not chosen? Why do you think their budget was so high? Why did the selection team require a budget to be included with the proposal at this early stage? Do you think budgets should have any bearing on the construction team member selection?

___

26.2   The above project was extremely successful in many regards, but like most projects, it had its ups and downs. The general contractor's original budget was slightly more than the client had anticipated. A six-week value engineering (VE) process ran in parallel to preparation of the design development documents. Some VE items were accepted and incorporated. The owner also ran a new proforma based upon other market conditions and increased the budget. After completion of the construction documents and development of a $22.5 million guaranteed maximum cost (GMC) contract, construction commenced. The following summarizes some of the project's statistics:

- There were 100 change order proposals (COP's). Some of these were for credits
- There were 250 requests for information (RFI's)
- The project finished three months behind the originally proposed schedule
- There were problems with two major subcontractors as discussed below
- There were not any outstanding liens or claims at close-out
- The final contract amount was incredibly close to the originally proposed budget
- The contractor did not make their full fee, but still realized a fair profit
- There were not any savings to share
- All of the team members are looking forward to working together again

Should every project conduct a VE process, even if it is within budget? Why do you suppose the project was considered so successful? How could the budget and the final costs have been so close?

_____
_____
_____
_____
_____

26.3   Assume the opposite result. The client was very dissatisfied. Using materials presented above and below in this case, prepare an owner's claim, highlighting all of the project's "failures" rather than successes.

_____
_____
_____
_____
_____

26.4　One of the reasons this project ran late was because it had a slow start with a shoring subcontractor. The general contractor competitively bid all of the normal subcontract areas, including shoring. They chose the low bidder who was a large but out of state firm. This firm imported their people and equipment and also engaged several other second tier subcontractors for portions of the work. Early in this phase of the project, the owner's representative witnessed what appeared to be a very poorly managed operation. He mentioned this to the contractor's site superintendent. Was he right to do this? Should a client or designer become involved in a contractor's "means and methods?" What sort of liability issues might arise?

26.5　The superintendent's response to the owner's representative was that he was staying out of the subcontractor's management problems. Was he right to do so? Could he contractually force the subcontractor to change foremen, equipment, or second tier subcontractors?

26.6　The shoring subcontractor ultimately finished one month late. The general contractor was never able to recover from this delay. The subcontractor also submitted several unsubstantiated COP's six months after demobilization, all of which were rejected. What are the lessons learned here for all parties?

26.7. The other problem subcontractor on this project was responsible for the window wall system. Review the following chronology:

- Three individuals from an experienced glazing firm recently began their own firm and were the successful bidder for this project
- The subcontractor was not able to post a performance and payment bond but the GC pledged to watch them closely
- The subcontractor subcontracted (brokered) all of the work to third tier firms
- Submittals were turned in late
- They submitted on an alternate manufacturer. After extensive discussion of potential schedule impacts to use the specified manufacturer, and verbal guarantees from the GC, approval was given for the alternate
- The submittals were incomplete and required several revisions
- The subcontractor relied on a third party firm to verify field dimensions
- Materials arrived from another third-tier fabricator late
- The original third-tier installation subcontractor was changed out at the last minute
- The new installation subcontractor could not adequately staff the project
- The GC and the architect immediately questioned the quality of the installation
- Lack of proper submittals and accurate field dimensions failed to identify a discrepancy in the construction documents (whose fault is this?)
- The manufacturer, fabricator, and installer were not being paid by the prime glazing subcontractor/broker and the GC was writing two-party checks
- Yikes, what if this leaks?

What errors were made by the project team, which caused or failed to prevent each of the above issues?

_____
_____
_____
_____
_____

26.8 The general contractor (GC) stepped up and assumed direct control of the window wall package. The prime subcontractor was paid off and their contract was properly terminated. The GC then contracted direct with the manufacturer, fabricator, and installer. The direct management of these third (now second) tier firms by the GC was now much more intense. This portion of the project was finished with acceptable quality and within days of the original duration. Was the GC therefore successful in this area? What sort of risks did they assume when taking control of the work? Did they protect the client? Who ultimately provides the guarantee of the windows?

SECTION 3: CONTRACTS

Including Insurance and Bond Issues

Case	Title
27.	Moving Target
28.	Budget or Bid?
29.	Turn-Key Impasse
30.	Historic Restoration
31.	Residential Dispute
32.	Shared Savings
33.	Subcontract Bonds
34.	All Inclusive?
35.	Seismic Repairs
36.	Line Item Estimate
37.	Allowance Accounting
38.	Contract Concerns

Most of these case studies overlap with other primary topics and at least 24 other cases involve contracts. See Appendix 2 for a matrix connecting all 101 of the cases with all fourteen primary topics.

Case 27: MOVING TARGET

This large $400 mil public works project was competitively bid utilizing drawings which everyone thought were relatively complete but turned out were just a progress set. The contract format was to utilize a GMC rather than lump sum which would have been more customary for bid and public works projects. A fee of 4% was stated in the GMC. Shortly after the contractor started the project, a completely new document set was issued. The contractor was to continue with construction based upon these new documents and prepare a change order, or revised GMC in parallel. One month was allowed for re-estimating. Almost as the estimate was being finished a third completely new document set was issued. The client put aside the requirement for change order number one since this third set of documents was really the set which was going to be built to and hopefully incorporated via change order. The contractor then began to estimate these documents. This same process repeated monthly throughout the two years while the project was under construction. A total of 20 complete sets of revised documents were issued. The contractor continued on with the project and kept very accurate accounting records. The quality of the work, schedule adherence, and safety awareness were all managed well. The contractor eventually turned in a very large change order, an increase of almost 50% over the original "bid," backed up with "actual" costs, not estimated costs. Aren't actuals more accurate than estimates? Was this a lump sum project? Was it a conventional GMC project? Or, was it actually completed as a cost plus percentage fee project? Was this fair to the taxpayers? Many public projects end up with the contractor claiming the owner. Do you think that happened in this case?

Case 28: BUDGET OR BID?

A residential client has hired a general contractor to construct a major $3 million dollar executive home. The client contracted separately with a reputable architect for the design. After the permit was obtained the architect's contract is closed out. The contract authored by the contractor discusses in detail what is reimbursable but does not tie the GC to a fixed price. The $3 mil is referred to as a "budget" in the contract. The owner had assumed because of verbal discussions with the contractor prior to the contract execution, and because of estimates provided by the GC on paper where the word "bid" was used, that this was a lump sum $3 million agreement. These discussions and estimates were not tied into the contract. There was not a third party owner's representative or agency CM involved and the owner had never been involved in a construction project. During the course of construction the owner and the city had requested several changes but none of them were formalized into the contract. Many of these were additive but some were also deductive changes. Many of the changes were due to building code changes. Because the general contractor felt this was a T & M project, they had not felt it was necessary to submit change orders against a "budgeted" amount. The GC had previously invoiced and received 67% of the original budget, or $2 mil from the owner. These invoices were all reviewed and approved by the lender. At approximately 90% completion the GC now invoices the owner, but for the full 100% amount of the original $3 million. When pressed, the GC indicates that the project will over-run the budget by approximately $500k. The owner and the GC disagree. The GC pulls off the job and refuses to do any more work until the owner agrees to the revised budget of $3.5 mil and pays the $1 mil now due. The owner dismisses the GC at this point. Both parties sue each other. How did this happen? Who is at fault? What should have been done to prevent this situation? What do they do now?

_____

_____

_____

_____

_____

Case 29:   TURN-KEY IMPASSE

29.1   A privately held small industrial manufacturer has come to an impasse with their general contractor. The client is a first time builder and engaged the contractor to design, permit, and build a relatively simple $5 million pre-engineered steel building. There was neither an architect nor a professional owner's representative on the project. Most of the space is to be open span with a bridge crane running full length. The crane will be contracted separately direct from the manufacturer to the client. The GC planned bare bone offices and restrooms in an attached structure. The dispute is in regards to the level of finishes and a mezzanine in this attachment. The GC's CEO owned the local franchise on this type of facility. He provided the drawings and obtained the permit and therefore sees himself as the interpreter of the documents. Is he correct? The GC has refused to perform this questionable scope unless they receive a change order for $50,000. Can an owner force a contractor to proceed with work that is in dispute? The client stopped the current progress payment of $200,000 for other work already completed. Can they do this? The contractor subsequently stopped all work and demobilized. Can they do this? Is this a good tactic?

29.2 A mediator was brought on board. When the mediator asked to review the contract, the client indicated that the contractor never provided one. The contractor responded that there was a contract and produced the original two page proposal which was signed "received" by the client. The client had not been provided with an original or copy of this proposal after signing. Without any further formal agreements, is this signed proposal the contract? The conditions of this proposal were consistent with the contractor's actions. When is a turn-key contract beneficial to each of the parties?

29.3 Take the position of the contractor and make your five-point case to the mediator.

29.4 Take the position of the industrial manufacturer and make your five-point case to the mediator.

29.5 What recommendations will the mediator make to each of the two parties?

Case 30: HISTORIC RESTORATION

The following facts are available for the dispute resolution parties to evaluate:

a. This is a historical restoration (substantially different than a remodel or renovation) project. The home is over 100 years old and is listed on the national historical register. The owner (and also occupant during renovation) is a medical doctor.

b. The owner has employed a one-person architectural office. The architect is licensed and she is reported to have this type of restoration experience. She is a single mother with a toddler. She has her child with her at most meetings with the owner, contractor, suppliers, and the city.

c. The architect recommends the owner engage a general contractor who is reported to have completed similar projects in San Francisco. There is not any documented resume or brochure from the contractor in evidence that supports either that he made these representations or that he has this experience.

d. The owner of the construction firm also works as a carpenter and the superintendent and project manager on his projects. His annual volume is approximately $200k. This was his sole project for the two years he works on it. He invoices his time as 'cost'.

e. The architect provides a one-page list of work that "may" need to be performed during the restoration project. There are not any drawings or specifications. The list uses many terms such as "if needed," "as necessary," and "if requested by the owner." Foundation repair is not one of the items listed. Most of the items on the list relate to finishes such as millwork, light fixtures, and wallpaper. This list is not referenced or attached to the contract.

f. The GC provides to the owner a one-page budget of $700k for the work. The owner reports that this is over his budget and they ultimately agree on a budget of $600k. The one page budget is marked up in hand with the revisions to achieve this figure. Neither party initials the revisions. The revised budget is not dated and also not attached to the contract. All of the figures on the budget are round figures. The contractor reportedly did not obtain outside subcontractor or supplier input or performed any QTO.

g. The GC authors a contract to the owner. This is a "home-grown" document. There are not any general conditions. Neither the budget figure nor the budget document, nor a schedule are referenced or attached to the agreement. The contract specifies the hourly wages to be paid to the craftsmen. It turns out later that the actual wages paid to the craftsmen are less than those specified in the contract, although the contractor bills according to the contract. The contract specifies that the GC will be paid 15% overhead and 10% profit on top of the cost of the work. Terms such as "overhead, profit, reimbursable, or cost" are not defined. Neither the owner nor the contractor obtained legal assistance to review the contract.

h. The contract is silent with respect to change issues, schedule, dispute resolution, subcontract practices, and retention. The contract requires the GC to "bid out subcontract work," although subcontract is not defined.

i. The contractor begins the project without any drawings or specifications or a building permit. None of this is finalized during the entire two years the contractor is employed. The contractor obtains several minor "subject-to-field-inspection" permits for small portions of the work.

j. The architect only produces rough 8 ½ x 11 sketches, all without reference numbers or issue dates. There are not any directional documents such as CCD's or CCA's provided. The best documentation available (after deposition) from the architect is her personal log. The bulk of the documentation later available from the contractor are photographs of actual site conditions and construction progress.

k. The owner continues to make changes and add scope to the work during construction such as an under-ground wine cellar, swimming pool, and an under-ground garage. Many substantial floor plan changes are made. All of the existing plumbing and electrical and insulation is determined to be inadequate and requires replacement. Significant foundation and roof structure damage is discovered and requires extensive rework.

l. Many unknown conditions are discovered such as solid bedrock under the foundation which must be jackhammered and removed by hand. The existing foundation system has failed and lacks any reinforcement steel and must be replaced. The entire four-story house must be jacked up to accommodate this work. The sanitary sewer pipes are completely plugged and some of the craftspeople are infected with giardia from working around raw sewage.

m. There were not any written requests for change orders originated or executed by any of the three parties during the two years.

n. During deposition, none of the original three primary parties are able to produce any written documentation such as memos, diaries, letters, phone logs, change orders, meeting notes, field questions, submittals, etc.

o. The contractor invoices the owner every two weeks and is paid promptly until the last pay request.

p. The contractor employs many subcontractors during the process, authorizing all of them to proceed on a T & M basis and signing the subcontractors' proposals rather than issuing subcontract agreements. This also applies to suppliers and purchase orders. The subcontractors later make these documents available to the court. The general contractor evidently did not retain copies. The Doctor claims the GC was contractually required to obtain lump sum subcontractor bid quotes.

q. The owner replaces the first architect about 16 months into the contractor's two-year employment. The second architect is also a one-person shop, but he documents to the other extreme. On an average day he will write 5 to 10 formal letters to many of the parties involved. Many of these letters use very direct and often adversarial language.

r. During this second architect's tenure on the project, which lasts just shortly after the original general contractor is dismissed, he continually complains in writing about the lack of definitive direction from the owner and the lack of as-built field conditions or available materials from the contractor. The lack of this information impacts his ability to finalize the drawings.

s. The owner contracts with all the second tier design firms direct. Many of these firms are rotated in and out of the project. There are not any second tier design firms that start on the project who also finish the project.

t. The owner sometimes dictates to the contractor which subcontractors to use, but the subcontractors' money is run through the GC's books, all with markups. The owner will, from time to time, fire a subcontractor, also without going through the GC. The owner will communicate direct with the subcontractors on change of scope issues. Very few of these directions are documented, except the occasional reverse memo from a subcontractor.

u. There are not any liens filed by any of the designers, subcontractors, or suppliers.

v. The original general contractor is dismissed after two years of employment. He has reportedly spent $1.2 million. He has completed approximately 10% of the actual work that eventually will be required to finish the home according to the Doctor's intentions. Many of the interior finish items on the first architect's scope list were not even started by the original GC during his tenure. The GC does not receive payment on his last $60k invoice. The GC liens the project.

w. Two years after dismissing the first GC, the Doctor files a lawsuit. He claims the GC had an obligation to perform all of the necessary work and achieve the original budget. He claims the GC miss-reported his accounting data. The Doctor claims the contractor did not have a right to charge his own wages to the job, as he was the owner of the construction firm. The GC also charged a 2% bookkeeping fee for his wife's accounting efforts for accounts payable and accounts receivable. The Doctor does not feel the wife's bookkeeping charges are cost reimbursable. The Doctor claims that some of the work accomplished does not meet code and other work was of such poor quality that it required rework by subsequent contractors. The owner's claim is for $400k.

x. The owner will eventually go through two more GC's and one more architect. As of this writing, more than four years after the project was started, it is still under construction.

30.1 Argue the contractor's case to the arbitration panel using at least three project management tools. Use facts from both within and outside of this case and your classes. One team will win and one team will lose in arbitration. Be creative. Anticipate the owner's response in your preparation. You only get one chance to present your case. You will not get a rebuttal.

30.2   Argue the Doctor's case to the arbitration panel using at least three project management tools. You only get one chance to present your case. Use facts from both within and outside of this case and your classes. Be creative.

30.3   As a third party arbitration panel, what would your judgment/decision be? How much is the award and to whom? Take a firm position. Base your decision on the information you have learned in your text, classes, these presentations, and the "documents". Be right. Your credibility, and your score, will be completely lost if the instructor "in the robe" over-turns your decision. Sell the arbitration method to the class. Why is arbitration better than mediation for this specific case?

30.4   Take the position of a court of law. Assume that the arbitration process in 30.3 was "non-binding" and the losing party appeals the case to you. Do you uphold or over-turn the arbitration panel's findings? Substantiate your decision. Why is court better than arbitration, mediation, or a DRB for a case such as this?

Case 31:   RESIDENTIAL DISPUTE

The following facts have been presented by these two opposing parties for the dispute resolution teams to decide upon.

a. This is a dispute between a general contractor and a client on a custom home project.
b. The owner owned the land prior to the agreement.
c. The GC authored the contract, which was a homegrown document, i.e. non-copyrighted.
d. The GC borrowed some of the AIA A201 general conditions items, but not all, did not incorporate nor reference the AIA document, and modified those which he did copy.
e. There wasn't a third party owner's representation from a CM firm nor did the architect provide these services.
f. The architect was from out of town and did not have any prior relationship to either party. The architect's involvement was discontinued at the time of issuance of the building permit.
g. The contract does not indicate it was "cost plus" but this was the builder's intent.
h. The contract does not indicate it was "lump sum" but this was the client's intent.
i. The project was negotiated between the two parties. There were not any other competitive bids received.
j. There was only one change made to the original agreement by either party and it was inserted by the owner and agreed to by the contractor. It indicated that "The only items which are not in the contractor's control which can result in changes to the contract amount are increases in the purchase cost of materials and scope increases originated by the owner. These items, if exceeding the contract amount, will be performed on a time and material basis."
k. The original estimate by the contractor was $650,000. This was value-engineered down to a contract amount of $580,000.
l. The value-engineering (VE) changes included shelling some of the interior space, reducing the square footage of the house from 5300 SF to 4400 SF, and (verbally according to the GC) reducing some of the finish and appliance estimates. None of the VE items were recorded on paper nor incorporated into the contract.
m. The contract references the original permitted and estimated drawings.
n. There are not any finish or appliance specifications in any of the documents.
o. During the course of the project the client's wife would choose finishes and appliances from the builder's recommended suppliers.
p. According to the builder, he advised the client's wife that they were overspending the estimate. None of this is documented. There was not any signed change orders to the original agreement.

q. There is not any discussion in the documents or in the contract regarding allowances for finishes and appliances.

r. The builder finished the house and has invoiced the client for approximately 10% over the contract amount for a total of $640,000. The client has paid up to the contract amount of $580,000.

s. The city issued a C of O and the client has moved in.

t. The contractor liened the home for $60,000. This lien has prohibited the owner from obtaining permanent financing. The owner is still paying on his construction loan, which is now 2% above current market rates. The bank is pressuring the owner to obtain permanent financing.

u. The GC paid all of the suppliers and subcontractors and none of them filed liens.

v. The GC has essentially been paid its "cost" but has not received any fee which coincidently amounts to approximately 10%.

w. The GC and the owner had worked successfully together on other professional endeavors. Both have good reputations in town. The client is an attorney by trade.

x. The contract mandates arbitration as the dispute resolution method.

y. The owner has filed a counter-suit due to damages associated with the lien and financing for $30,000

z. Both parties are also attempting to recover legal fees.

31.1 Argue the contractor's case to the arbitration panel using at least three project management tools. Use facts from both within and outside of your classes. One team will win and one team will lose in arbitration. Be creative. Anticipate the owner's response in your preparation. You only get one chance to present your case. You will not get a rebuttal.

31.2 Opposite to 31.1 above, argue the owner's case to the arbitration panel using at least three project management tools. You only get one chance to present your case. Use facts from both within and outside of your classes. Be creative.

31.3   As a third party arbitration panel, what would your judgment/decision be? How much is the award and to whom? Take a firm position. Base your decision on the information you have learned in your text, classes, presentations from the parties, and the documents. Be right. Your credibility, and your score, will be completely lost if the instructor "in the robe" over-turns your decision. Sell the arbitration method to the class. Why is arbitration better than mediation for this specific case?

31.4   As a third party mediator, how would you propose to bring the two parties together to resolve the dispute? What solution would you propose to each to resolve the issue? What is mediation? Sell the mediation method to the class. Use information you have learned from your text, classes, outside sources, and the presentations from the parties to base your proposal. Be creative. Why is mediation better than arbitration for this specific case?

31.5   What should the owner have done to keep this from happening? What should the contractor have done to keep this from happening? What should the bank have done to keep this from happening?

Case 32: SHARED SAVINGS

A general contracting (GC) firm and developer have successfully completed several projects together and have a solid working relationship. All contracts between the two have been negotiated Guaranteed Maximum Cost (GMC) contracts. Change orders on past projects have been minimal and the General Contractor has always been willing and able to use project savings to compensate. Instead of adding more estimated dollars to the GMC for change orders, the GC has absorbed the cost of the extras from their in-house buy-out fund, which would have been submitted for a split at the end of the project.

On the last project the GC was able to move 25% of the value of savings used to cover change orders, which was their portion of the savings split, into their fixed fee during the course of construction, rather than waiting for the end of the project. Both parties benefited from this arrangement.

Their current project is a $30,000,000 office building that is half way through the over-all schedule and is near completion of the structural work, which is where most of the GC's labor risk is. There have been a large number of agreed upon (i.e. not disputed) COP's for added scope and document discrepancies equaling approximately $500,000. It is still relatively early in the project and the project manager is confident that this time there will not be enough savings to compensate for all of these changes. The project manager has kept the developer notified of changes by submitting change order proposals (COPs) for the work and sharing his change order log. The developer has given the GC verbal direction to proceed with the changes but refuses to sign the COPs until the GC can present savings as on previous projects. The PM documents these directions in his weekly meeting notes. Now, with all of the subcontractors securely bought out and most of the estimating risk behind him, the Project Manager is expecting only approximately $100,000 in savings but remains reluctant to release it until the building is weather tight.

32.1   Take the position of the general contractor. Assume that the GC considers the developer a key client and does not want to jeopardize their relationship. What should the project manager do? Should they eat all the COPs as before?

32.2 Take the position of the developer. Can you contractually proceed in this fashion? How do your defend you position?

32.3 Where is the architect in all this? Take the position of the architect as a mediator and resolve this situation.

Case 33: SUBCONTRACT BONDS

Your firm has negotiated a 30-story high-rise office complex in downtown. This is a very major project both for your firm's volume and reputation. Unfortunately for you, as the PM, and your client, the market was very busy when the subcontracts were bid. Subcontractors with whom you do not have prior relationships or history will perform many of the major categories. Your guaranteed maximum cost (GMC) proposal and contract with your client requires that your firm post a 100% performance and payment bond. The cost of this bond was anticipated and is included. At the time of the execution of the bond, your bonding agency is requiring that you also bond all second tier subcontractors and suppliers whose values are greater than $40,000. This is a total of approximately $15 million worth of subcontracts. Their average bond price is 2%; therefore the bonds will cost approximately $300,000. This value was not included in your GMC estimate. You approach the client and ask them to pick up these fees, but they respectively decline. Should these firms be bonded? Is this standard that your bonding agency would require these bonds? Where did you error? What can you do now?

Case 34: ALL INCLUSIVE?

A developer and an architect bid out an $8 million mixed use facility (MXD) to three general contractors. The drawings are less than perfect, maybe 60% complete at best. One reason for bidding at this stage was the decision by the developer to hold down the architect's fee. None of the contractors asked any questions during the bid cycle. The documents include many phrases such as:

- Build according to industry standards
- Must meet all codes and city requirements
- The documents are not meant to show every detail
- The contractor is responsible to report discrepancies prior to contract award
- If there is a discrepancy, the more stringent or most expensive detail shall apply
- The architect is the sole interpreter of the documents and her interpretation shall stand

There are of course numerous discrepancies and resultant requests for additional funds. The contractor has submitted 50 requests for change orders and the structure is not yet complete. The developer and architect remain strong that the contractor should have assumed an answer in his estimate and should have included appropriate contingencies. All of the change orders are rejected. Half way through the project the contractor pulls off. The GC has been paid $3 million of the base contract but has not received anything for the $400,000 worth of change orders which are on the table. The GC has heard from their subcontractors that at least double this figure will be coming when they mobilize on the project. Can the contractor walk away from a project such as this? What sort of notice would be required? The owner will of course claim the contractor has improperly terminated the contract and they are being damaged for delay. Are "all inclusive" terms such as this fair in the documents? Assuming this case goes to court, how would a judge rule?

Case 35: SEISMIC REPAIRS

The owner of this multi-tenant retail facility is a 70 year old retired construction management professor. He has invested his life savings into the building and acts as his own property manager. The facility received significant cosmetic and structural damage during an earthquake. All of tenants were evacuated until repairs were complete. The owner immediately employed an architect, structural engineer, and a general contractor to implement repairs. The owner felt in this manner he was acting on the best interest of himself and his tenants. By expediting repairs, he would get his tenants back into the facility, thereby minimizing loss of sales. He felt he was also looking out for the insurance company. The design teams' corrections were well documented and the contractor's costs were tracked on a time and material basis. The insurance company also immediately mobilized but did not take any action. They preferred to sit back and watch the corrective work take place. Upon completion of this several million-dollar repair project, the insurance company then questioned whether the fixes were proper and whether the actual costs reported were accurate. They also pulled out a provision in the property insurance policy, which required the owner to receive competitive bids on all portions of the repairs. The insurance company ultimately offered the owner 80 cents on the dollar for repairs, which the owner passed through and offered to both his design and construction teams. Will they accept? Who loses in this situation? What should the different team members have done to have assured 100% collection? Is this a standard situation in the insurance claim arena? If the owner sues the insurance company will he collect 100% of costs incurred by himself and his agents? Will the tenants pitch in?

Case 36: LINE ITEM ESTIMATE

A software development client entered into an agreement with a general contractor to build a simple 50,000 square foot office building with an open office concept. The client's architect questioned three estimate line items in the contractor's guaranteed maximum cost proposal, which appeared to be heavy. The contractor responded that these costs may be needed and the owner will receive 90% of any savings according to the terms of the contract. The contractor did not manage the project efficiently and finished four months late. Because of this delay, the contractor reported that they over-ran the bottom line GMC and there were not any savings to share. If a contractor mismanages the work, does the owner lose an opportunity for savings? The architect closely monitored the work involved with the three subject estimate items and was confident that the contractor under-ran the estimate in these areas. The contractor indicated that it was not a "line item guaranteed maximum cost estimate, but rather only the bottom line was what was guaranteed." If there were savings on single line items, especially ones which were questioned early, can the contractor use these savings to offset over-runs in other areas? Will an audit now help either party? What do standard contracts say?

Case 37:   ALLOWANCE ACCOUNTING

An electronics manufacturing facility has employed a general contractor to construct a 1000 car post tension underground parking garage. There are three allowances included in the guaranteed maximum cost contract. All three are essentially contingency funds which the contractor may need due to rain delays, poor soils, and dewatering, each totaling $100,000. None of these conditions truly came to pass. When the client asked for the allowances to be change ordered out of the contract, the contractor delayed. They argued that the contract did not state when the allowances were required to be accounted for. How and when are allowances formalized into the contract? As time went on, the contractor used creative accounting and charged all sorts of marginal costs against the allowances. Ultimately the allowances were all 'spent'. How can allowances be defined and managed so that both parties are treated fairly? Should allowances exist? How are allowances any different than an open ended time and materials contract?

Case 38: CONTRACT CONCERNS

Your firm is competitively bidding a hotel project on the waterfront. The market is slow. You have experience in this geographic area and the hotel construction arena. Your firm has worked with the prime architect on numerous successful projects. The architect is contracted by the national hotel chain, which will operate the hotel when complete. Your contract will be with the property owner who is retaining ownership of the dirt which makes them a JV partner in the completed facility. The bid documents include a homegrown contract which appears to have been developed by the hotel's attorney. The contract references an attached AIA A201 general conditions document, which has been substantially modified. These two contract documents have several conflicts. The bid form requires your officer's signature acknowledging that you do not take any exceptions to the proposed contract agreement. You know the architect from previous projects and have an opportunity to review their contract with the hotel operator and discover that this is also in conflict with the construction contract documents. Do you send your contract to your attorney? What will be their advice? This is a bonded project. Do you send the contract to your bonding agency? What will be their advice? Do you pull out of the competition? Do you submit a qualification with your bid stating all of your contract concerns and recommended changes? What will your competition do? Do you stay silent to all of this and choose to fight it out after the award, or wait until an issue arises during the course of the project?

SECTION 4: ESTIMATES

Case	Title
39.	No Subcontractor Coverage
40.	Subcontractor Short List
41.	Unknown Electrician
42.	Plan Centers
43.	Estimating Too Smart
44.	Window Bids

Most of these case studies overlap with other primary topics and at least 24 other cases involve estimates. See Appendix 2 for a matrix connecting all 101 of the cases with all fourteen primary topics.

Case 39: NO SUBCONTRACTOR COVERAGE

A pharmaceutical firm is letting a project to bid to a select list of four qualified general contractors. The project will ultimately be awarded for $27 million, which is approximately $1000 per square foot. This is very complicated work. The architect is very experienced. The client's local west-coast office is utilizing the services of its corporate "owner's representative" from the east. This individual has a reputation of beating up the entire construction industry, including GCs, subcontractors, designers, and even the City. This will be the fourth project for this client locally and none of the local GCs have worked for them on a second project. Contractor 1 does not turn in a bid. Contractor 2 is involved in litigation with the owner from a previous project and the owner essentially disqualifies their bid. Contractor 3 violates the bid rules and is disqualified. Contractor 4 did not really want the project due to their busy workload and the client's combative reputation, but turned in a courtesy bid. Contractor 4 was surprised to receive a phone call from the client notifying them of intent to award. The estimator responsible did not pursue the subcontractor industry for quotations. Because of the client's reputation, many of the qualified subcontractors did not pursue the project. 30 different specification sections received one or zero quotations and were plugged by the estimator. Oh, did we forget to mention that contractor 4's estimator just quit?

39.1 What errors did contractor 4 make in the estimating process? Should they accept the award? Can they ethically go out and re-bid to the subcontractor industry? Are they in a strong or a weak position for buyout opportunities? Who has the upper hand with contract negotiations in this scenario, the contractor or the owner?

39.2 Because the GC did not anticipate getting the project, and the estimator wanted to distance himself from it as soon as possible, they had to go outside to hire a "contract" PM. This PM had a similar reputation to the Owner's Representative. He was known to be extremely tough and the GC felt this is the type of personality they needed to come out even, if not ahead. How do you think the two individuals approached doing business with each other? How did they deal with the other team members? Would you be surprised to find out that this project was a huge success?

Case 40: SUBCONTRACTOR SHORT LIST

An outside third-party project manager (AKA Owner's Representative) working for a public university solicits bids from three general contractors with the stipulation that the major subcontractors must be chosen from the client's pre-approved list. The successful general awarded the electrical scope to a firm which was not on the list. The client's representative is aware of this situation but remains quiet. The second-place general contractor finds out about this infraction and files a complaint. Should the university stay with the original award? Should the project be re-bid by GC #1 to the select electrical firms? Should the project be awarded to GC#2? Should the entire bid process be thrown out and started over? Is this fair to the taxpayers? Is it fair to the low electrical firm? Can either the low general contractor or the low electrical subcontractor file a complaint if they are not allowed to proceed with the project? Does this situation make a good case to disallow short listing on public works projects? What should the owner's representative have done to prevent this from happening?

Case 41: UNKNOWN ELECTRICIAN

For the last four weeks three individuals in your firm plus yourself have worked full time estimating a $50 million public bid hospital project. Your volume is very low and you need this project. You figure that your office is easily $25,000 in the hole already with estimating expenses. On bid day you receive an unsolicited bid from an out of state electrical subcontractor. Their price is $750,000 below the lowest local firm. You have never worked with this firm before and do not know anything about their qualifications. The local electrical inspector has a reputation of being very tough. With ten minutes to go before the bid is due your office calls the electrical firm but their phones are jammed and you cannot get through. Their emailed bid was generically addressed to your firm and simultaneously to your four major competitors. What will your competition do? What will you do?

Case 42: PLAN CENTERS

You are a staff estimator for an out of town general contractor. Your firm wishes to start a satellite office in this location but you would like to have a project going before you commit the resources for a move. The project you are bidding is a simple concrete tilt-up warehouse. You have met with both the client and the architect and are comfortable that they will treat you fair. The architect has filed the documents with all of the internet plan centers. The general contractors you are competing with are local firms who have a subcontractor base from which to work. You must rely on unknown subcontractor prices developed from the plan centers. There are four addenda issued during the bid cycle. You have received the addenda and know that they are substantial and cover a lot of disciplines. The plan centers do not have any fiscal or contractual responsibility to make sure that all of the subcontractors have seen all of the addenda. You complain to the architect but his response is that all of the other general contractors seem to be dealing with the situation just fine. On bid day many of the competitive subcontractor bids which you receive do not acknowledge receipt of all four of the addenda; many of them do not acknowledge even the first addendum. You end up with the low bid, but submit a qualifier about the problem with the addenda. The owner and architect indicate you must take the job as is and that all four addenda will be tied into your contract. Is this fair? Do you have any recourse? What can be done about the plan centers to rectify this type of situation? Will you accept the project with these conditions? How can an out-of-town contractor compete? What would be the best way to open a branch office?

Case 43: ESTIMATING TOO SMART

You are the owner's representative for a new electrical trade's educational facility. At the request of the Board of Trustees you have put this relatively simple $2 million project to bid to a short list of four general contractors. You are very familiar with three of the GCs who are of comparable size and experience. The fourth is an out of town firm who is substantially smaller, but has constructed these types of facilities in the past. Two of the larger general contractors are relatively silent throughout the bid cycle. They diligently pick up their drawings, attend the pre-bid meeting, and appear to be earnestly estimating the project. The third larger firm asks over 100 detailed questions about the documents.

The estimator is a friend and you advise him that he is becoming "too smart" on the job. All of his questions are answered in writing and issued as addenda. The smaller contractor does not attend the pre-bid meeting. His estimator calls you with only ten days left before the bid is due and wonders where he can pick up additional drawings and when the bid will be due. He does not ask any detailed questions. The morning the bid is due he asks for a time extension but is denied. As anticipated, the detailed estimator turns in the high bid. The two silent contractors are in the middle. The smaller general turns in a bid that is $100,000 lower than the others, but is within 5% of the total. The client is very delighted about the tight bid results and with the low bidder. What should your advice be? Is it ethical to disqualify someone who was on a short list? If hired, what will eventually happen with this low firm? Is this fair to the contractors who prepared what is anticipated to be "complete" bids? Which firm will prepare the fewest change orders?

_____
_____
_____
_____

Case 44: WINDOW BIDS

You have received three window-wall subcontract bids. The low bid of $500k was from an unsolicited subcontractor. They did not fill out your prescribed bid forms. They neither acknowledge nor deny the bid documents or addenda. You know nothing about this firm. Their bid does not state any exceptions or qualifications regarding their proposal. The second bid of $600k is from a glass firm which you have worked with prior on other sites, but not successfully. The quality of their work was fine, but they battled with your office with respect to contract issues. They were short-listed and requested to bid on this project. They filled out your bid form, but their proposal came with an extensive list of exclusions, assumptions, and qualifications. The third bid was also from a glazing firm whom you had solicited a proposal from. They are a local firm but you have not worked with them prior. Their bid is exactly per your prescribed bid form. They do not state any peripheral qualifications or exceptions. Their bid is $700k. Your pre-bid budget was $600k for this area of work. What do you do? How would your answers differ if this were a) a lump sum competitively bid job, or b) a negotiated GMP job?

_____
_____
_____
_____

SECTION 5: SCHEDULES

Case	Title
45.	Glazing Schedule
46.	Drywall Subcontractor
47.	Liquidated Damages
48.	Schedule Hold

Most of these case studies overlap with other primary topics and at least 9 other cases involve schedules. See Appendix 2 for a matrix connecting all 101 of the cases with all fourteen primary topics.

Case 45: GLAZING SCHEDULE

You, as the GC's PM, have a problem glazing subcontractor. They are behind schedule. They refuse to work overtime to catch up. The subcontractor has submitted several unsubstantiated change order proposals (COPs) that have not yet been approved and they are threatening to stop work. They have switched out both the project manager and the superintendent since the project started. They are not staffing the project according to their planned and committed manpower. You are not getting along personally with the subcontractor's current project manager and have resorted to communicating only through email. You are receiving pressure from the field to resolve the problem and get the glazier to perform. Your supervisor has indicated that it is your responsibility to solve the problem. What do you do? What could you have done to prevent these problems from occurring? What recommendations can you make to a general contractor's subcontractor management system to prevent these types of situations?

Case 46: DRYWALL SUBCONTRACTOR

Your drywall subcontractor is not performing in the field. They have not staffed the project according to their original commitment or to your field superintendent's expectations. They are holding up the work of other trades. The quality of the drywaller's completed work has been unacceptable and you are constantly on them to improve. You have asked for the removal of their superintendent (do you have this right?) but the firm has refused (can they do this?). It is eventually decided by your home office that the subcontractor must be terminated. How do you go about this process? Is it simple? How is it documented? Will you get sued for false termination? What does standard contract language say? How do you protect yourself? Will it be easier to just keep limping along with them?

Case 47: LIQUIDATED DAMAGES

Your client has assigned $2000 per day liquidated damages (LDs) for late completion to your $25 million turn-key contract. You in turn have passed these liabilities on to your subcontractors. You have one siding subcontractor who sends you written notices requesting additional time whenever an RFI or a submittal is a day late being returned. They document every adverse weather day. They document when other subcontractors are holding them up. They request schedule extensions with every change order. Not all of their documentation is substantiated, but some of it may be. They are claiming a total of 20 additional workdays. This is an administrative nightmare for you.

47.1   How do you deal with the siding contractor during the course of the project? Did you pass through these notifications to your client? If this subcontractor ultimately finishes behind schedule by ten days, and you also finish behind schedule, does your client collect from you? Do you collect from the subcontractor? How does it get resolved typically? What is the contractual and legal resolution? Analyze both ways: a) with stipulated liquidated damages in the subcontract agreement, and b) without.

47.2   Do you offer a subcontractor a bonus ($/day) for finishing early? Isn't this only fair? If the penalty were $2000 per day for finishing late, what would the bonus be? Assume the siding contractor in 47.1 above finished per their original schedule, but had built up the claimed additional 20 days, and had such a bonus clause. They would now claim that a bonus was due for the $40k. Do they get it? If not, why not? If so, who pays; you or your client? How is this resolved?

Case 48:   SCHEDULE HOLD

48.1   Your firm does $200 mil in annual volume with a home office overhead of 2%. This particular 18 month project is worth $20 mil and there are 8% jobsite general conditions. Assume that at exactly the mid point of the schedule your project was put on hold for one month due to reasons beyond the contractor's control such as weather, union strikes, owner financing, or city issues. Pick one. Using the contract, classes, your text, and research outside of the classroom, how should the contractor properly deal with this delay? Discuss issues such as notice, documentation, jobsite administration costs, home office costs, loss of fee, loss of productivity, and quality and safety concerns. What is the 'Eichleay Formula'? Prepare a claim for this delay.

48.2   As an alternative to 48.1, prepare a recovery plan to get the owner their building on time. Show with the schedule that a recovery is possible. Submit a change order proposal for the anticipated costs associated with working in an expedited fashion. Provide all necessary cost backup.

48.3   Similar to 48.1 but assume this time that the fault is with you, the general contractor through the actions of your site utility subcontractor. The underground fire loop was installed and back-filled without requesting inspection from the city. The city subsequently required the entire system to be re-exposed, which seriously impacted all of the work on the project due to lack of access. From the general contractor's perspective, how do you deal with the subcontractor? How does this impact other subcontractors?

48.4 Given the scenario in 48.3 above, how do you respond to the impact this has caused the owner? From the owner's perspective how do you deal with the real costs that the delay described in case 48.3 has placed upon you? Do you have any recourse contractually? Compare the differences and philosophies (advantages and disadvantages for all parties) between LDs and Actual or Real Damages

SECTION 6: SUBCONTRACTORS

Including Subcontracts, Purchase Orders, and Suppliers

Case	Title
49.	HVAC Union
50.	Union General Contractors
51.	Hospital Buyout
52.	Young Engineer
53.	Carpet Bankruptcy
54.	Steel Supplier
55.	First Team
56.	Concrete Walls
57.	Pulled Quote
58.	Hostile Project Manager
59.	Fire Protection Heads

Most of these case studies overlap with other primary topics and at least 31 other cases involve subcontractors and suppliers. See Appendix 2 for a matrix connecting all 101 of the cases with all fourteen primary topics.

Case 49: HVAC UNION

This general contractor has chosen a union Heating Ventilation and Air Conditioning (HVAC) subcontractor on a predominantly open shop project. The HVAC subcontractor is very qualified, is the low bidder, and so far is performing well. But now: First, some of the crew is damaging roof flashing and louvers, which are under contract of the open shop roofing subcontractor. Then, they intentionally slow down their own work which is slowing down work of other trades. Eventually they pulled off the job. The union sheet metal trade is picketing the job and bothering those who cross the picket lines. These actions are all from the crew, not the HVAC subcontractor – this is costing them money. How did this happen? Who in the general contractor's organization is responsible for solving the problem? Can you contractually force the subcontractor to replace the crew and/or return to work? How can you prevent this from happening in the future?

Case 50: UNION GENERAL CONTRACTORS

Many general contractors are signatory to the carpenter and laborer unions. A few others also have agreements with trades such as cement finishers, ironworkers, operating engineers, and teamsters. If a contractor is signatory to one, but not all unions, are they considered to be a "union contractor?" Are union GC's required to employ subcontractors such as electrical or plumbing who are also union affiliates? Do union subcontractors feel that a union GC should only employ all union subcontractors? What is customary? Do union subcontractors only work for union GC's? Why the difference? Does a carpenter-signatory general contractor only employ union drywall subcontractors (who are also carpenters)? What would be the rule if the union general contractor subcontracts work out which they normal self-perform, such as horizontal formwork for elevated slabs performed by carpenters? Can they do this open shop? Do market conditions or geographic locations affect any of this?

Case 51: HOSPITAL BUYOUT

During buyout of a $40 million hospital expansion project you as the GC's PM are faced with the following possibilities:

a. You can package three specification sections (drywall, insulation, and paint) all under one $2 mil subcontract. This will all be work that the sub will self perform. 1) If the packaging results in a slight cost increase of $10k, does this reduce your work effort and is it worth it? 2) Conversely, if the packaging results in an overall cost savings of $50k to the general contractor, how can you be assured that you are not over-loading the subcontractor? 3) And finally, if it is a push in cost is it a good idea or a bad idea and why?

b. You received a bid from a qualified concrete reinforcement steel (re-bar) installation subcontractor for a portion of the work which your firm normally self-performs with your own ironworkers. It is $100k less than you have estimated. 1) List 3 reasons why you should hire this sub. 2) List 3 reasons why you should not hire this sub.

c. Assuming that you did hire a subcontractor for work which you normally self-perform, you are now faced with stiff criticism from your field supervision because you have taken work away from "his guys." The sub is not performing to your superintendent's satisfaction. He is on your back continually, requesting you to terminate the sub. Why might you have thought this was a good idea? How do you now deal with this issue?

d. Even though this is a lump sum project, your client has requested that you solicit a bid from and do your best to employ his favorite painter. Should you include the subcontractor on the list? If the sub is second bidder by $25k, and the owner agrees to pay the difference, should you hire them? Is this ethical? If you proceed and the painter does not perform, what role, if any, does the client share?

e. You have a bid from a floor-covering subcontractor who is a "broker." They will buy the material and will hire a separate subcontractor for installation. The subcontractor's price of $400k is competitive. Is this to your advantage? Why or why not? What is your firm's position to this arrangement? How do you keep the prime sub from stepping back and allowing you to manage their third tier subcontractors and suppliers?

f. You have an opportunity to employ a subcontractor who will buy the doors, door hardware, and doorframes from three different suppliers, something you normally purchase direct, and have them installed all in a $175k package deal. This will be $7,500 more than if you purchase the materials separately and hire an installer. Is this to your advantage? Why? At what premium is it not in your advantage? Draw an organization chart to reflect this scenario. Show contractual as well as lines of communication.

_____
_____
_____
_____
_____

g. If you were not allowed to package the door system in 51.f above, what can you do to assure that all of the work will be performed properly? If the doors and frames do not come out machined properly, who is at fault? Who checks the shop drawings and the submittals? Who pays? Is there a back-charge procedure to follow? How do you solve the problem in the field? Draw an organization chart to reflect this scenario.

_____
_____
_____
_____
_____

h. You have an opportunity to break apart a $4 million mechanical bid proposal that is normally purchased as a package. The prime HVAC sub has offered you this opportunity to save his mark-up of $300k on managing these multi-layered subcontractors. This includes separate fire protection, controls, mechanical insulation, plumbing, and pipe labeling subcontractors. Take the position that you will accept this deal. Your supervisor recommended against it, but allowed you the rope to succeed. Why did you feel this was to the general contractor's advantage? How will you manage the work so that you personally succeed? If it becomes a problem during the project, how can you correct it? Draw an organization chart for this scenario.

i. Take the opposite position to 51.h above. Draw an organization chart packaging all of the mechanical trades under the HVAC subcontractor. Is this the correct arrangement? Do they manage their sub-trades adequately or will you have to? At what savings is it worth splitting this work apart?

Case 52: YOUNG ENGINEER

You are responsible for two subcontractors, earthwork and roofing, on your mega-project. You are a 23-year-old field engineer. Experienced 40 and 50 year old project managers work for both of these subcontracting firms. Both of these subcontractors are very qualified and have worked with your firm on previous occasions. The firms are financially strong and their fieldwork is of acceptable quality. Your problem: the Rodney Dangerfield syndrome, "you don't get any respect." These subcontractors continually go over your head to your supervisor. They talk to your peers on other projects. They do not respond to your requests for information or requests for additional backup on change orders. What do you do now? Do you care? How do you establish yourself as the point of contact for these firms? List five rules of order you would recommend for a beginning engineer who needs to earn the respect of the subcontractors who work for him.

Case 53: CARPET BANKRUPTCY

Your firm has not worked with this $500k carpet firm before. You, as the GCs PM, originally decided to bond them but were surprised that their bond rate was 4% of contract value. They have submitted several change order proposals for both increases in scope and discrepant documentation. Most of these are pending processing due to one reason or another. The quality of their work is fine. They are supporting your schedule. You have recently received multiple material men's notices and pre-lien notices from the carpet subcontractor's suppliers. The subcontractor has indicated that they are solvent and do not have any cash flow problems. You have just discovered that the chief executive officer of their firm has left. You have made a few calls and have caught wind that they may be going bankrupt. What do you do? What contractual recourse do you have? Should you hold up future payments? Should you notify your client? Will the third-tier suppliers come after you for payment? Can they? Assuming the subcontractor sells their receivables to a collection agency will they also come after you? What adjustments to the system can you recommend to prevent this situation from happening in the future?

Case 54: STEEL SUPPLIER

Your firm has never worked with this structural steel fabricator prior. They have been marketing your staff purchasing agent and he has decided to give them a try and issued them a purchase order and handed it off to you as the GC's PM – gee thanks! This is a very complicated detailing project. The fabricator has subcontracted the detailing out. There are numerous questions and meetings with the detailer. The shop drawings show up relatively on schedule but require a substantial amount of your time for review. There are numerous changes that are brought up during the shop drawing process. Many of these will ultimately result in justifiable change orders to the client. In short, you are totally engulfed in the detailing effort. The fabricator has invoiced you for detailing, as well as the complete mill order material purchase and some fabrication. You have paid accordingly. You had no reason to suspect that anything was wrong. The embedded steel members show up on time and have been fabricated properly. You have not visited the fabricator's shop. Your first major shipment of columns and beams is scheduled for Thursday. On Monday your supervisor discovered that you had not inspected the steel in fabrication and suggested you take a trip to the shop. On Tuesday you show up at the shop unannounced and cannot find any of your steel. The fabricator tells you it is temporarily being stored at the rolling mill. You visit the mill and find out that not only is the fabricated steel not there, but the mill has not received a purchase order for your project. In fact the mill will not do business with your fabricator due to lack of payment on a previous project. But, the mill has the shapes and lengths you are looking for in-stock. On Wednesday you receive a bankruptcy notice from your fabricator.

54.1 How did this happen? What should you have done to prevent it from happening? Who in your organization is at fault?

54.2 Do you try to keep the fabricator afloat? Do you put your people in their shop to fabricate the iron? Do you take on their personnel and pay them directly?

54.3 You decide to hold any additional payments that are in the system for the steel fabricator. There is some money due them for detailing and embeds. The supplier and detailer both file liens on your project. Are they valid? How do you get them removed? Your client discovers the situation and begins to pressure you to have the liens removed. Can the client contractually withhold all further payments from the GC until this is resolved? How could you have prevented this from happening?

54.4   How do you get the steel delivered to the project and still keep your schedule afloat? Do you re-bid? Do you negotiate with the second bidding firm? Assume that the market has gotten tighter and all of the fabricators have plenty of work. Do you hire the second bidder lump sum, T & M, or unit price?

54.5   Assuming you find another fabricator. Will they want to re-detail the project? Will they accept the detailing which already exists? Is this a good idea for you? What are your risks either way? How do you mitigate these risks?

54.6   Your other subcontractors catch wind of this problem and begin sending you notices of potential impact costs associated with delays. How do you deal with these subcontractors? Are you up front or do you try to hide the problem?

Case 55: FIRST TEAM

A client has selected a general contractor to perform pre-construction services and to eventually enter into a negotiated contract to build a $20 million shelled office building. The construction market is very busy. Many good subcontractors are turning away opportunities to do work. The people who are now managing (office) and supervising (field) construction projects in the subcontract arena are not typically the first team selections. During a busy market many PEs are promoted to PMs and foremen to superintendents. Predict how this project could turn out if the general contractor posts the project in public plan centers to receive open market bidding from subcontractors and suppliers. As a general contractor it is your duty to protect the client from these risks. Prepare a list of several project management tools which could be used to mitigate both the general contractor's and the client's risks.

Case 56: CONCRETE WALLS

Your firm was selected as the GC on a ten-story office building project. Your contract includes many generic terms such as "contractors", "work" and "documents". It does not differentiate between a GC and their subcontractors or specific items of work. You bid out the concrete package and have three tight bids within three percent of each other. The scope includes form work, concrete placement, and finishing. You also estimated the concrete work yourself and were within three percent of the low bidder, just pennies different from the high bidding subcontractor. Before awarding to the apparent low bidder you decide to interview them to make sure they have the scope and schedule covered. Their bid did not reference any drawings. In the interview you ask if they have all concrete shown on both the architectural and structural drawings. You asked this question because there were a couple of housekeeping pads and one set of steel stair in-fills shown on the architectural drawings which you wanted to make sure were picked up. They answer yes to the question. This was documented in the pre-award meeting notes. These notes were not made an exhibit to the subcontract.

You decide to hire that subcontractor for the concrete work. One week later you mail them a subcontract and at the same time they start the work. Two weeks later you get a call from the subcontractor inquiring why the reference was made to the architectural drawings. You remind the PM of the pre-award meeting and the concern that he picks up the housekeeping pads and stair in-fills. He continues working. Two weeks later they sign and mail back the subcontract but cross out (properly initialed) the reference to "concrete work shown on architectural drawings." As it turns out, there are several concrete walls that are also shown on the architectural drawings but are not shown on the structural drawings. The walls are not detailed as they would have been on the structural drawings, which would typically include steel embeds and rebar. A field question is written and the structural engineer responds with the necessary sketches, but the architect does not attach a directional document such as a CCD to the field question response and the sketches. You forward this information to the subcontractor, directing them to proceed, and again remind them of the pre-construction discussion you had, including another copy of the meeting notes.

56.1 The concrete subcontractor refuses to accept the subcontract as originally written and will not install these walls unless a change order is issued. The walls were not specifically discussed in this prior meeting. They continue working. The walls are worth about $50,000. You do not have the money in your budget to take care of the issue and neither does the subcontractor. What can you do? What went wrong and how could it have been prevented? If this were a lump sum project would this be a change order to the owner? Would it be accepted?

_____
_____
_____
_____
_____

56.2 Assume this is a GMC job. One condition of the owner's selection was that you could not self-perform any work unless you received competitive bids and were the successful bidder. What would have happened if the GC had been awarded this package to self-perform this work? Would the GC have asked for a change order if they did not have the dollars in the estimate? Could the GC make a case for a COP if they could cover the $50k from other savings in the estimate? Is this a good example of why a negotiated GC should be allowed to self-perform their normal work tasks?

56.3 What can designers do to anticipate and mitigate these types of situations? Should design firms package work according to standard subcontract or labor jurisdiction lines? Should Owner-GC contracts make hard distinctions between the General Contractor and Subcontractors and not use the term 'contractor'?

56.4 Who ultimately pays for these walls? Argue your case to the arbitration board assuming one of these positions:

a. Concrete Subcontractor
b. General Contractor
c. Structural Engineer
d. Architect
e. Owner

Case 57: PULLED QUOTE

A metal decking supplier submitted a $300,000 bid to the negotiated general contractor on a large distribution/warehouse facility. The second supplier's bid was $600,000. After minor value engineering revisions the general contractor requested a subsequent bid from the low supplier which was then even lower than the original bid. After further strong-arm negotiations from the general, they finally settled on a price of $250,000. The GC finalized their negotiations with the client based upon this figure. One month later the supplier declared that they had a bid error and was pulling their bid. There was not a bid bond between the general and the supplier and they had not yet signed their purchase order (PO). The GC had negotiated and executed their contract with the client. The general contractor made a plea to the client to be allowed to raise the bid by $350,000 so that they could sign up the second bidder. The client denied. The general contractor took this case to court. The lower court agreed with the client, and ruled that 1) the GC should have known the low bid was in error, and 2) the GC should not have worked the low bidder down. On appeal, the upper court reversed this decision and agreed with the general contractor's claim. Their reasoning was:

1) There was no way for the general contractor to have known that maybe the higher bidder was in error
2) The GC has proven bid error
3) The client would have had to pay the additional $350k originally if the low bidder had either estimated correctly or not submitted a bid. The client was not due a windfall benefit

How did the low supplier error in their post bid actions? How did the GC error? What could the owner have done to prevent this from happening? Do you agree with the first ruling? Is the second ruling correct? What would be the fairest way to resolve this situation for all parties?

Case 58:   HOSTILE PROJECT MANAGER

You are an experienced project manager working for a flooring subcontractor. Your younger counterpart on the general contractor's side is hostile to all of his subcontractors. He challenges your RFI's, returning many without forwarding them to the client. He routinely cuts your change order pricing and your pay request draws without consulting you. You decide to deal with this individual by not bringing problems to his attention. You proceed "per plans and specifications" as he has directed you on numerous occasions. You have worked with the GC's CEO on previous occasions. He is a people person and encourages cooperation and open communication at every opportunity. He has told you specifically that he is looking for long-term relationships with subcontractors who are part of the "team-build" solution. You have told him directly that his project manager is a problem and the CEO has responded that he will look into it and appreciates your dedication to this project. The general contractor project manager has found out that you have discussed this situation with his supervisor. Things are now worse than ever. Are you in the wrong? How do you deal with this situation today? How could you have dealt with it differently from day one? Once you have gone up the ladder and over someone's head with a problem can you ever go back down for a solution?

Case 59: FIRE PROTECTION HEADS

You are within two weeks of turnover of a three story build-to-suit office building project. Many of the city inspections are complete and it appears that a certificate of occupancy (C of O) will be obtained on time. Only the fire protection, elevator, and life-safety remain. The mechanical, electrical, fire protection, and life-safety were all design-build systems. During the punchlist it is discovered that the fire protection subcontractor has installed a different type of sprinkler head on each floor. The first floor uses concealed heads with white escutcheons to match the ceiling tiles. The second floor uses decorative chrome heads. The subcontractor installed semi-recessed heads on the third floor. This was not picked up prior as a person cannot physically see more than one floor at a time. The client is very upset. Can you require the subcontractor to change them all out now? Would this be a good move given the status of the inspections? Who is responsible to check and approve the shop drawings and submittals for design-build subcontractors? What should the general contractor have done to prevent this from happening? What can you do now?

SECTION 7: STARTUP

Including Pre-Construction, Mobilization, and Value Engineering

Case	Title
60.	No Pre-Construction Agreement
61.	20% Value Engineering
62.	Missed Pre-Construction Meeting
63.	Early Mobilization
64.	Formal Value Engineering
65.	Subcontractor Value Engineering
66.	Private Value Engineering

Most of these case studies overlap with other primary topics and at least 3 other cases involve startup. See Appendix 2 for a matrix connecting all 101 of the cases with all fourteen primary topics.

Case 60: NO PRE-CONSTRUCTION AGREEMENT

You are the project manager for a general construction firm and you think that you have successfully sold your services to a client who is building a golf course clubhouse. You have been performing informal pre-construction services for three months. This includes budgets, schedules, value engineering, meetings, meeting notes, and constructability analysis. You did not have a formal signed agreement for your services. You later discover that the designer has a contract to complete 100% plans and specifications. You also discover that the client intends to bid the project out competitively. It is rumored that your client may have been receiving pre-construction input from another general contractor in parallel with your efforts. You have spent $20,000 already on pre-construction services. Did you make an error? If so, how did you error? What do you do now? Why do GCs offer pre-construction services below cost and sometimes even for free?

Case 61: 20% VALUE ENGINEERING

The owner has reported that this general contractor's GMC is over their budget and has asked for value engineering (VE) ideas. The owner needs to cut over 20% out of the estimate. Is this possible? Should contractors volunteer to lead the VE process? Do architects embrace this process? How should the owner and contractor deal with potential revision items that the architect does not embrace? Is the VE process "cheapening" the project? What liabilities has the contractor assumed by proposing changes? How do VE changes get incorporated into the contract?

Case 62:   MISSED PRE-CONSTRUCTION MEETING

A utility subcontractor receives a permit to connect a new branch water line to a main in a heavily traveled street. The subcontractor is working for a general contractor. They shut the line down at 7 a.m. and begin tearing up the street. The city shows up at 9 a.m. and shuts the project down because the contractor did not hold the prescribed pre-construction conference. The street is now cut up to the centerline, a 6-foot deep shored trench exists, and the water main has been exposed. Obviously the error is the contractor's (but which one?) and the answer to the question: what did they do wrong....is easy. But now is the city correct in shutting the project down? Morning rush hour traffic was obviously impacted but is now over. The afternoon rush hour traffic will begin at approximately 3 p.m. The continued shut down of the water line will impact fire service and water service to the local neighborhood. What should the course of action be? Should the general or subcontractor be fined? Should the general contractor or subcontractor be eliminated from bidding on future projects? In the public arena, can a contractor be eliminated from bid lists? Whose responsibility was it to schedule the pre-construction meeting?

_____
_____
_____
_____
_____

Case 63:   EARLY MOBILIZATION

A general contractor had previously submitted a budget of $14 million to the owner of a medical office building (MOB) project. The budget was based on 70% design documents. The contractor was selected based on the experience of its project team and its approach to the project. The contractor had been given a pre-construction services contract for $20,000 to join the owner's project delivery team. The contractor participated during the balance of the design phase of the project, performing value engineering and constructability analysis, and providing input to the construction drawings. During the last two weeks of the preconstruction phase, the general contractor mobilized on the job site. The owner neither directed nor stopped them from doing so. The terms of the contract had not yet been finalized. The contractor set up the site camp, brought temporary utilities to the site, and began the initial surveys and layout. The construction drawings were issued and incorporated all of the team's input, along with the city permit comments. The general contractor then prepared a $15 mil Guaranteed Maximum Cost (GMC) estimate based upon these revised documents, which exceeded the previous budget by $1 mil. The owner was extremely upset and would not listen to explanations or reasoning why the estimated costs increased. The general contractor was asked to move off of the site. When the GC requested to recover the additional $15,000 they had incurred for the two weeks on the site, the owner refused payment. Should the general contractor have mobilized on to the site? Why would they have been motivated to mobilize without a contract? Does the GC have any recourse for payment? What steps would you suggest for both the owner and the general contractor now? If the GC was entering into a lump sum contract, would their actions have been different and why?

Case 64: FORMAL VALUE ENGINEERING

Many new public education construction projects are required to conduct formal value engineering (VE) studies. This occurs early in the design development phase. It lasts approximately one to two weeks and employs several outside consultants, none of which will ever work on the project. The cost of this study may be in the $50,000 to $60,000 range. Most of the ideas put forth by the team are totally without merit. Many of the VE ideas are for just a few hundred dollars. Rarely are any of the proposals accepted into the design, yet all of the owner and designer team members are satisfied that they complied with the intent of the law. Is this true value engineering? Is it fair to the taxpayers? What should be the prescribed process? Should a percent of the construction budget be required to be either proposed as value engineering ideas and another percent accepted into the design? Should a third party employed by the state supervise the process? Should there be financial incentives?

Case 65: SUBCONRACTOR VALUE ENGINEERING

What are GCs approaches to value engineering? Do you request it from subcontractors? If subcontractors offer unsolicited cost savings ideas do you consider them? How do you assure that you are getting good value in return? Who owns the design responsibility for a value engineered item? Should you give the subcontractor some incentive to propose these ideas? Is it acceptable to allow them to skim some of the savings off?

Case 66: PRIVATE VALUE ENGINEERING

Value engineering (VE) was a significant element of the pre-construction services provided by this general construction management team on a highly complicated medical facility. The negotiated GC competitively bid out all subcontracted areas but the $66 million GMC was still ten percent over the client's budget; six million dollars would eventually be saved through the VE process.

66.1 A few of the VE proposals included the same scope but just a different method, material, or manufacturer, but most of the savings involved elimination of scope, shelling areas, reduction of redundancies, eliminating expansion capabilities, and simply providing a product of lessor value or shortened life. Is this true VE?

66.2 Should the GC receive a portion of VE savings? Should there be some fee enhancement? Is there a potential for fee reduction? What incentive does the GC have to reduce the cost of the project?

66.3   After the client has accepted the $6 million of savings, the design team must now undertake a significant document revision. Who pays for the redesign costs? If the architect provided the original $60 million budget figure, does this change the scenario? Which documents are 'contracted' to? Is this issue covered in standard AIA contracts?

66.4   The GC and the low-bidding subcontractors proposed almost all of the savings ideas. The design team formally went on record disagreeing with many of the ideas (is this typical?), yet the owner still accepted them. One mechanical proposal involved a different control valve. The valve never performed properly. The designer did not help with analyzing the problem or the correction. The owner blamed the GC and the mechanical subcontractor who proposed this $150,000 savings, even though the owner accepted it. Are contractors liable for performance of VE proposals? Are they assuming design risk? If so, why would they propose any solutions?

66.5   Do contractors like VE? Do they give back a dollar for a dollar's worth of savings? VE can also be for more expensive items and increase the GMC. If this were the case, do contractors pad these increases in estimated costs? Is VE a way for contractors to improve profits or cover for other estimate line item shortcomings?

SECTION 8: COMMUNICATIONS

**Including Documents, Documentation, RFIs, Submittals,
Written and Verbal Communication Issues**

Case	Title
67.	RFI Value
68.	Dropped Baton
69.	Architect's Administration
70.	Three General Contractor Project Managers
71.	Two Architectural Project Managers

Most of these case studies overlap with other primary topics and at least 17 other cases involve communications and document control. See Appendix 2 for a matrix connecting all 101 of the cases with all fourteen primary topics.

Case 67: RFI VALUE

67.1 A high technology client and their architect engage in the design and construction of a multi-phased multi-building campus. The total design and construction process will take over ten years. The buildings will all be similar in design and function, but not exact. The design is complex and the first building will eventually cost approximately $500 per square foot. A GC is selected for Building One-Phase One after a competitive proposal. It is their hope to build out the entire campus. The general contractor's project manager and his project engineers generated 2500 requests for information (RFI) RFIs, and 500 change order proposals (COPs) were subsequently negotiated in this first phase. The contractor pointed to document discrepancies for most of the issues. In addition, there were over 100 submittals that were completely rejected by the design team. The GC and the subcontractors also pointed to the specification ambiguities for this situation. There were not any outstanding claims, liens, or quality questions at the phase one completion. The job was completed on schedule and there were not any safety incidents. This was a successful project, wasn't it?

67.2 Ultimately the client engaged another general contractor for the second phase. The architect later boasted to the first GC PM that his successor only wrote 1000 RFI's and less than 200 COP's on that phase. Very few submittals required complete rejection. What happened? Was the second GC team easier and more "client friendly?" What did the design team do with all of the first GC's documentation?

67.3   The second GC continued on and built five more buildings and the remainder of the campus. The above RFI, COP, and submittal figures reportedly were reduced with each subsequent phase. Did the first PM do a bad job? What does this say about RFIs and submittals with respect to the design and the closeout processes? Aren't RFIs and submittals actually part of the design completion and early 'active' quality control? Who benefited from this process? Who paid? How could the first PM have kept his firm on the campus to construct the balance of the phases?

Case 68:   DROPPED BATON

The GC you work for has just relocated you from one job site to another. The Project Engineer (PE) you are replacing was "relieved of his command." He was very experienced and had a very strong personality. You will be managing the work of the HVAC subcontractor, although your prior experience was all related to civil and structural work. You don't know the difference from pipe and duct. You find that your predecessor's desk is completely empty. You review the files for the HVAC subcontractor and find them very brief, disorganized, and unprofessional. No one at your new site has any knowledge of the work that your predecessor or the HVAC subcontractor were doing, other than they fished a lot together. You cannot find an executed copy of the subcontract agreement. The invoices and change orders for this subcontractor are not even close to tracking. Upon notifying the subcontractor that you will be handling their account their PM indicates that he is very displeased about your predecessor's removal. "They had an understanding, and commitments were made." What do you do? What could be done to the general contractor's systems to prevent this "dropping of the baton"?

Case 69: ARCHITECT'S ADMINISTRATION

This project is a $4 mil elementary school remodel that was bid lump sum. There are numerous conflicts in the documents, many of which are associated with matching new to existing work. The project manager for the general construction firm has become frustrated with the lack of paperwork from the architect. The architect appears to have run out of construction administration funds. Some of the problems and responses to request for support are listed below.

- Responses to the RFI process include: "I don't want a written question, just give me a call"
- The architect will answer written RFI's with a verbal
- He will not meet in the field and review actual conditions
- Written responses quite often just indicate "see the plans or specifications"
- Submittals are late being returned and they often do not include any disposition
- The architect misses many weekly construction meetings, showing up at some late and leaving others early
- He never brings his meeting notes and does not acknowledge ever receiving them
- He is also not reviewing change orders in a prompt fashion

What should the project manager do to resolve this issue? What could the school have done to prevent this from happening? What risks do both the general contractor and the school incur if this situation is left unchecked? What risks does the architectural firm incur?

Case 70: THREE GENERAL CONTRACTOR PROJECT MANAGERS

The construction market is very busy. This general contractor is having a difficult time obtaining good qualified project mangers. The officer-in-charge (OIC) has hired three new project managers (A, B, and C) within the last few months. The OIC performed a diligent interview and reference check on each of these employees. He is just about to receive the following reports from his superintendents who are working with these individuals on three separate projects.

A. This project manager likes the detail. He is very thorough in his research and his work is very accurate. Unfortunately all of this accuracy is taking too much time. He tends to hold RFI's and submittals and subcontractor change order requests too long. He has a tendency to hand out last week's meeting notes at the beginning of the next week's meeting. Delay and impact notices are now coming in from all of the subcontractors.

B. This individual has a field background. He was a journeyman carpenter before going to college to get a construction management degree. He spends most of his day outside of the trailer. He is prone to provide direction to the GC's and the subcontractor's field forces, bypassing his superintendent and that of the subcontractors. His paperwork is being performed adequately during off-hours. The general contractor's superintendent is about ready to quit.

C. This PM was educated as a structural engineer. She has a professional engineering license and lets everyone know it. She is constantly second-guessing the design documents and design team direction, but she is always correct! She has written negative letters to the design team, and is copying the city and the client with these letters. The design team is about ready to mutiny; even a subcontractor has written a letter requesting to be let out of their subcontract. Her project management paperwork is excellent and timely. She returns a profit!

What should the officer in charge do? What will happen if these scenarios are left unchecked? If, due to a slow-down in the market, a) who would be the first PM the OIC would let go, or b) which one has the most potential to develop into a project manager? Which type of PM are you?

Case 71: TWO ARCHITECTURAL PROJECT MANAGERS

As a project manager employed by a general contractor you find yourself working with two different and difficult architectural project managers from the same firm on different projects. Your firm has a very good relationship with this architectural firm. Your CEO has told you to "deal with it" and not to damage your firm's reputation, at the same time make a fair profit. Assuming that you cannot change these professionals' personalities and you have six months to go to finish each project, how do you deal with each of the individuals as described below?

Architect A is young and has a very strong personality. He develops and distributes his own copies of meeting notes. He publishes his own field question, submittal, and change order proposal logs. He refuses to address your logs or notes in the meetings. His records consistently slant in his favor with respect to content, responsibility, and dates. He chairs any meeting he attends and demands that he sit at the head of the table. Change order proposals that you originate do not show up on his logs. He authored your contract and your AIA change orders. Many of your "issues" continue to be sidestepped.

Architect B is very experienced but appears at times to be laid back. She appears to be tired and over-worked. She does not take any notes during the meeting and continues to show up unprepared, forgetting her copies to important documents. She leaves meetings early for "prior commitments." She loses field questions, submittals, and pay requests. She does not recall verbal or phone conversations, and does not acknowledge receipt of emails and faxes.

SECTION 9: PAY REQUESTS

Including Liens and Lien Releases

Case	Title
72.	First Time Developer
73.	Subcontractor Over-Billed
74.	Supplier's Lien
75.	Shoring Retention
76.	Which Site to Lien?

Most of these case studies overlap with other primary topics and at least 14 other cases involve pay requests. See Appendix 2 for a matrix connecting all 101 of the cases with all fourteen primary topics.

Case 72: FIRST TIME DEVELOPER

This Realtor was envious of real estate developers and the "reported" large profits they make. He decided to try has hand in this area although he did not have any design or construction experience. With insider knowledge he came across a piece of property that was available from another developer which had a design and permit in hand for a twenty unit condominium complex. The first time developer dismissed the architect and acted as his own owner's representative.

The deal also came with a general contractor who had worked on pre-construction for the previous developer. This new developer negotiated a lump sum agreement with the same general contractor. The contractor authored the contract and proceeded with the project. The contractor was actually more of a construction manager than a general contractor, as is often the case with residential work. They subcontracted out all of the work and did not have a full time superintendent on the project. The developer was not introduced to any of the subcontractors and he did not have a written subcontractor list. The contractor did not post any bonds.

The residential market was good and the developer had 80% of the condominiums pre-sold (cash pending completion) even though the project was just drying-in (roof on, siding on, windows in) and ready for mechanical and electrical rough-in. All relations had been proceeding fine with the contractor. The quality of the work was adequate, there were very few change orders, and there had not been any safety incidents. The construction schedule was only 12 months long and the contractor was on, or even ahead, of schedule.

The new developer had obtained a loan for all of the construction costs and leveraged other personal holdings against the land. Progress payments of $2 million had been paid on time. The pay request schedule of values (SOV) was only 5 items long. There were not any conditional or unconditional lien releases submitted by the subcontractors or the general contractor. Retention was not being withheld from the GC. It turns out the owner of the construction firm was having an extra marital affair with his treasurer. For the entire course of the project she had not paid any of the subcontractors. Without any notice, the two of them skipped town and disappeared to the Caribbean and were never heard from again.

The subcontractors all immediately demobilized and filed liens against the property. They had all properly filed their material men's notices before the project began but the new developer was inexperienced in lien management and did not retain any of these notices. None of the subcontractors would return until the developer had paid them their share of the work in place, which was now approximately $2.5 million. The bank begins to pressure the developer. Many of the buyers begin to look for other residences. Some begin legal proceedings against the developer. Now what? List at least ten project management errors this developer made.

___

Case 73: SUBCONTRACTOR OVER-BILLED

After a non-performing drywall subcontractor has been terminated it is discovered that they have over-billed your general construction firm and you, as the project engineer, have over-paid them. Their contract was for $1million. The subcontractor billed and you paid $750k. You had originally received competitive bids and the subcontractor was barely low. The drywaller had been submitting a ten line item schedule of values with the pay request and you were sharing it with your superintendent to verify level of completeness. You now bid out the remaining work, and receive very competitive bids, all near $500k to finish the work. You are $250k short. Your officer-in-charge jumps down your throat for allowing this over-billing. You have been very busy, managing ten subcontractors on this job. How did this happen? Who is at fault? What can be done next time to assure it doesn't happen again? How do you solve this specific instance? Deduct $250k from your salary?

___

Case 74: SUPPLIER'S LIEN

You feel that you practice very pro-active lien prevention procedures. On this project you received material men's notices from your fire protection subcontractor. Each month you received proper conditional lien releases. At the completion of the job you received an unconditional lien release and exchanged it for the fire protection subcontractor's retention check. One month later you received word that a lien has been filed by a third tier piping supplier from out of state for material they supplied to a piping fabricator for your subcontractor. You did not know this third tier supplier existed. You paid the fire protection subcontractor, the subcontractor paid their pipe fabricator, but the fabricator did not pay the pipe supplier. The amount in question is $40,000. How did this happen? How can it be resolved most easily? What is the correct and legal resolution? Is the lien valid? Can the supplier legally remove the pipe? What does the client do now? How are liens removed? What should you do in the future to prevent this from happening?

Case 75: SHORING RETENTION

This GC hires a subcontractor to install a conventional temporary shoring wall for an underground parking garage. The subcontract has an inclusion that calls for the subcontractor's retention to be released within 30 days after fulfilling all of its contractual requirements. The GC's contract with the owner only allows early release of retention with owner's approval. One month after all shoring work was complete; the GC submits the normal pay application to the owner, but includes a request to release all of the shoring subcontractor's retention. The owner refuses to release retention for the shoring wall until the parking garage structural work is complete (which is another 60 days) and is holding up the entire pay request. What should the General Contractor do? Should the owner release the retention? If the owner continues to refuse, what recourse does the subcontractor have with the GC? What recourse does the shoring subcontractor have with the owner? If they lien the property should the GC pay the subcontractor out of their own funds? What position does the lender take on early retention release? Is early retention release (for select subcontractors) a good practice (for any or all parties) and why? What should have been done to prevent this situation from occurring?

Case 76: WHICH SITE TO LIEN?

This client and general contractor were under contract for two entirely different projects on two different sites. Some, but not all, of the subcontractors were the same. The work on the first project appeared to have been completed according to expectations. Six months after the retention was released on project one, the owner discovers numerous cracks in the concrete and the repairs appear to be extensive. The exact cause is unknown and will take months to determine. In the mean time the owner holds back any further payments on the second project. Can they do this? The subcontractors and contractor will lien, but which property do they lien against? If the owner cannot withhold funds, what recourse do they have to remedy project one? Can the contractor stop work on project two? Is this wise?

SECTION 10: COST CONTROL

Case	Title
77.	General Contractor Cash Flow
78.	Cost Over-Runs
79.	Low Forecast
80.	Residential Developer

Most of these case studies overlap with other primary topics and at least 7 other cases involve cost control. See Appendix 2 for a matrix connecting all 101 of the cases with all fourteen primary topics.

Case 77:   GENERAL CONTRACTOR CASH FLOW

You are a new project manager for a small general construction firm whose annual volume is $100 million. All of the work is proceeding well on your first project, which is a local retail facility. You have been preparing the pay requests and picking up the checks from your client. You know that your firm is being paid on your project. You have been approving the subcontractor and supplier invoices. You begin to hear rumors that your firm may be in financial difficulty, which is not because of you or your project. Your subcontractors have been complaining that they are not getting paid. When you check with your comptroller she indicates that the subcontractors' checks are "in the mail." Your client also begins to hear complaints from your subcontractors and pre-lien notices are being filed. What do you do?

Case 78:   COST OVER-RUNS

78.1   Three months into a lump sum project this project manager has realized that they are over-running costs on half of the direct work activities. They are under-running on the other half. The bottom line looks okay, maybe even on the plus side. Is this possible? Can he do anything about the codes that are over-running? Since the bottom line looks okay, should he even worry about it? His superintendent does not want to report any single line-item cost over-runs to the home office and asks you to forecast the original estimate in each case. How can the PM cover up this situation? What are the ramifications if he does what the superintendent is asking? As long as the project makes the original fee, does it really matter how individual cost codes turn out?

78.2   This is a similar situation to 78.1, except all of the codes appear to be over-running their estimates. The project looks like it is going in the tank to a tune of approximately $500k. What are some of the reasons this overrun could be occurring? Should the project manager and the superintendent just hold on and ride it out? What are some potential resolutions you could recommend to the team? Should they ask for assistance from the home office? Should they start looking for a new employer?

Case 79:   LOW FORECAST

A general contractor's project manager negotiated a cost plus fixed fee (CPFF) contract with an apartment developer but without a GMP. The contractor's scope was limited to utilities, site work, and foundations for this hillside project. There were extensive concrete foundation walls and shoring systems required in the design in order to place all eight wood-frame buildings on this steep site. The superintendent and project manager calculated the amount of fill necessary to place behind the retaining walls and stored that fill on the site during excavation and concrete work. The balance of the excavation was off-hauled. The dirt was clean structural fill. All was proceeding well with the project. The developer asked the PM to share his forecast against the original $5 million budget. The PM anticipated an under-run of $100,000 at this time compared to the budget. The developer then used this $100,000 to upgrade his kitchen equipment specifications. It turns out that the construction team under-estimated the amount of fill necessary for the walls and had to purchase select fill to be brought back to the site. The anticipated budget under-run was eaten up by this extra dirt. The developer claimed the contractor was negligent by hauling off needed materials earlier in the project and the contractor was liable for the $100,000 they had now spent. Is the developer correct? Should a PM share the forecast with a client? Is he required to contractually on an open-book job? Will this cause the PM to be more conservative with his forecasts in the future?

Case 80:   RESIDENTIAL DEVELOPER

A speculative residential sole-proprietor builder/developer began his career with one to two houses at a time. He gradually worked up to five and ten lot mini-developments. He had a good product and was pricing his homes competitively. The builder had established a good reputation with his banker, Realtor, and the subcontractor industry. He had always made his payments on time. He had an opportunity to step up and he purchased a 50-lot sub division. The land was priced 20% high but the market was hot and he figured he could make a big killing. Most of the land cost was leveraged. He built and sold 5 houses without a problem. His Realtor was encouraging him to step up production and keep the prices high. He decided to proceed with the next 20 all at the same time, which was not his normal business model.

About half way through framing interest rates jumped two points and the residential market came to a screeching halt. The developer continued with the 20 houses but was only able to move two of them, and they went for 10% below asking price. The remaining 18 sat for six months with no activity. The Realtor was now suggesting the developer cut his prices dramatically. He dropped his price to below cost and was able to sell eight more. The last 10 were held for another six months. All of his creditors were filing liens. The bank had made several extensions on the land and construction loans and was now threatening foreclosure. The developer was able to sell the second 25 lots to another developer but they went for 30% less than the original purchase cost, which ultimately put him further in debt without any possibility for recovery.

This first-time developer filed for bankruptcy. His reputation was tarnished in the small community his family lived in and they eventually moved to another state, attempting to run from criticism and the creditors. How did the developer error? What mistakes did the subcontractors make? Should the subcontractors have put community and relations first and held off on the liens? Should the bank have loaned him for the original land? Should the bank have also held off on foreclosing? What should the bank have done to prevent the developer from becoming so extended? Was the Realtor at fault?

SECTION 11:   QUALITY CONTROL

Case	Title
81.	Brick Detail
82.	Subcontractor Quality Control
83.	Carpenter Classrooms
84.	Bid-Design-Build
85.	Sinking Swimming Pool
86.	Brick Mockup
87.	HVAC Units
88.	Wrong Carpet

Most of these case studies overlap with other primary topics and at least 4 other cases involve quality control. See Appendix 2 for a matrix connecting all 101 of the cases with all fourteen primary topics.

Case 81: BRICK DETAIL

This architect is very new to her career. The firm she is working for is quite small and specializes in educational facilities. She has been assigned to a public university project, which is more than double the value than either she or her firm have ever undertaken. This firm is very much a "per plans and specifications" architectural firm. For example she will return submittals 100% rejected just because five copies were forwarded, not the six which were required in the specifications. The general contractor and the general contractor's project manager who are working on the project have just the opposite background than that of the design team. The construction team has mostly private negotiated project experience, and they are very accustomed to large projects. They are the largest commercial contractor in the three states which they work. The project manager and the architect are not working well together.

81.1   How would you predict this project will conclude? What can the GC's PM do to adjust to this type of system?

81.2   On one site visit, this same architect noticed the exposed interior brick lintels were not what she had intended. There were not any clear details for this work. She did not discuss it with the general contractor but rather returned to her office and discussed with her supervisor. Two weeks later at a weekly construction coordination meeting she reported to the owner's representative about this deviation and stated she wanted it added to the quality control (QC) issues log. The brick mason foreman had chosen to repeat another lintel detail that was available in the documents, although it was for an exterior wall. This soldier course layout he followed was actually more difficult than the detail which the architectural firm desired. The foreman thought he was doing the right thing. What reaction will the general contractor's project manager and superintendent have during the meeting? Did the subcontractor error? How should field QC issues be reported and to whom and when? What will the owner's representative do?

Case 82: SUBCONTRACTOR QUALITY CONTROL

This general contractor was constructing a very sensitive medical tenant improvement (TI) project. The superintendent was very laid back and the project manager (PM) was at times combative with the client and the architect. The architect noted two quality control concerns early in the project that warranted more attention. This was the preparation and flashing of roof penetrations and the taping and fireproofing of the vertical drywall surfaces. These conditions occurred around a procedure room that required close attention of all the team members. The superintendent's response to the drywall issues was "the subcontractor is the expert and I don't tell him what to do." The PM further responded to the roof issues with "unless you want to pay me a change order to change the process, we will stay with our course. If it leaks, the roofing subcontractor will have to return and incur the cost of repairs." Is this a "client friendly" GC? Are they contractually correct? The architect is trying to implement active QC by bringing potential problems up early. Is this a good practice? Given the GC's response, will the architect continue with early notifications? If this is a "means and methods" issue, can the architect force changes without a change order? Can she issue an AIA CCD to force the issue? If the architect's concerns go uncorrected but eventually turn out to be a warrantee issue, is the owner in a better position to claim impact against the general? Will the owner or architect choose this GC again?

Case 83: CARPENTER CLASSROOMS

This project includes new classrooms and training workshops for the carpenters' union apprentices. The general contractor has bid the project lump sum and is approximately 90% through the schedule. GWB is being taped and finishes begin to arrive on the job. On a recent walk-through the owner and the design teams realize that several items are not exactly as they had anticipated. The gray carpet is actually black. The plywood wainscot in the warehouse area is CDX (construction) grade, not AC (finish) grade. The design-build sanitary waste pipe in the ceiling space between the two floors is ABS (plastic) and not cast iron. It minimally meets code and will be noisy. The gates on the fence are swinging and not rolling. The interior wood trim is hemlock and not oak. There are many other examples of these types of surprises. The contract requirement for preparing submittals was generic and although it did list a few items to be submitted, it did not list everything. There were conflicts in the documents. The contractor has chosen the most inexpensive materials wherever possible, and is now basing their argument on document inconsistency. The architect is recognizing that there were inconsistencies, but is pleading to the contractor that if there were questions, they were to ask with an RFI, or a submittal could have been used to verify their intentions. It is now too late to make changes and the owner will have to "live-with" these conditions. Using these or other specific material examples, show how project management tools could have been used to prevent these surprises. How could each of the parties improve their performance? Is this a way to achieve repeat negotiated work for any of the parties? If there had been a withholding of funds by the client, who would win the dispute?

Case 84:   BID-DESIGN-BUILD

This general contractor has been awarded a two-story speculative office building project on a lump sum bid basis. All of the MEP systems work, including HVAC, plumbing, electrical, and fire protection were also bid on a lump-sum but design-build basis, with very little criteria information available for them to base their bids on. The subcontractors were responsible for preparing their own documents and having them stamped by a state licensed engineer and obtaining their own permits. The general contractor did not submit these documents to the owner for approval, as it was not specified that they had to do so. The systems subcontractors have routinely received city inspections and approvals for work in place. As the project nears completion, the owner and his architect have just walked the project and they have discovered that several areas of the design-build subcontractors' work is not up to their expectations. This includes:

- HVAC: The ceiling is being used as a return air plenum. This is much less expensive than utilizing a ducted return air system. It is also noisier, less efficient, dirtier and requires plenum-rated cabling.

- Plumbing: The bathroom plumbing fixtures appear to be more residential than commercial grade. They are less expensive but they do meet code.

- Electrical: The light fixtures being installed are 2 x 4 prismatic versus more energy efficient, more expensive, and more attractive deep cell parabolic.

- Fire protection: The fire protection sprinkler heads have not been installed center of ceiling tile and are not lined up in the large open office areas and hallways.

The owner is now withholding a current pay request for $300,000 requesting the contractor to correct what he feels are deficiencies. Can he do this? Should the general contractor keep proceeding? Should the GC force the subcontractors to fix the problems? Do subcontractors care about client satisfaction? How would your answer differ form a negotiated versus lump sum bid project? How will this be resolved? What could the contractor have done to keep this from occurring?

Case 85: SINKING SWIMMING POOL

Maybe you should start this one by developing a timeline. 20 years ago this Doctor had a swimming pool built on a hillside adjacent to their residence by a local reputable general contractor (GC1). The Doctor was confined to a wheel chair. Because of this, he did not inspect the supports under the pool upon completion of the original construction. Ten years later, during a rainstorm, the hillside supporting the swimming pool suffered a slide that may have affected the pool supports but no one performed an inspection at that time. The pool continued to function as originally constructed.

A Geotechnical evaluation was directed five years after the slide as part of a re-financing package. The major support for the cantilever of the pool was three feet below the pool and was not providing any structural support. The study concluded that, although the swimming pool was constructed properly and had performed well, additional foundation stabilization through a pin-pile technique was recommended. The Doctor followed the Geotechnical engineer's (E1) recommendation and hired a contractor (GC2) to complete the required work. This second contractor was given all of the recommendations and specifications from the engineering evaluation and was contracted to properly complete the modifications. The contractor invoiced and was paid for work completed. Neither the owner, bank, nor the engineer inspected the repairs.

Following an earthquake three years later, the Doctor contacted a third contractor (GC3) to complete a visual inspection of the swimming pool structure. The contractor inspected the pool and discovered poor workmanship and potentially faulty construction methods used by the second contractor. Contractor number two apparently failed to positively attach the piles to the supporting beam, relying on gravity and weight instead. Contractor three approached an experienced Geotechnical/concrete expert to assist with the project. Upon inspection, this second engineer (E2) concurred that the repairs made three years earlier by contractor two were not according to the first soils expert's (E1) recommendation. They now called for new concrete and soils testing to see if the supports could be re-attached with a new bracket that could hold it in place and provide equal support as originally intended.

85.1 Who should pay for the additional work required to correct the current problem? Is it the original contractor (GC1) who built the pool almost 20 years prior? Is it the second contractor (GC2) who inadequately installed the added supports after the slide? Is it either the first Geotechnical engineer (E1) or the bank which didn't inspect the repairs after the slide?

85.2 Should the homeowner's insurance policy pay for all of the repairs following the recent earthquake? Is the homeowner responsible because he did not adequately inspect the work, and in effect by paying the earlier parties, let them off the hook? Is the latest team of contractor (GC3) and engineer (E2) just looking for more work? Do inspecting contractors and designers and financiers have an incentive to find fault with existing conditions? Are they liable if they inspect and do not uncover defects? Is any of the work under warrantee? Does this qualify as a "latent defect?"

Case 86: BRICK MOCKUP

The architect for this brick facility had specified the requirement for a full-scale mockup of all the exterior closure elements around a window. This involved four different subcontractors and approximately 10 different types of materials. Are mock-ups an active or passive quality control technique? Are they submittals? How long should a mockup remain intact on the site? The general contractor on this project attempted to value engineer the mockup out of the project. Is this an example of active or passive quality control? Why would they want to delete the mockup? The brick mason used their two best foremen to build the mockup. Was this a good idea? Who should build mockups? Later, both of the foremen moved on to other projects. The subcontractor had a difficult time repeating the level of quality established in the mockup. The architect prepared a punch list before the mockup was accepted so that the contractor could continue work. Was the designer's intent for this mockup fulfilled on this project?

Case 87: HVAC UNITS

The general contractor on this project chose not to purchase the Heating Ventilation and Air Condition (HVAC) equipment from the mechanical subcontractor but rather they purchased them direct from the manufacturer. The shop drawings were submitted and were per specifications. The submittal was approved and returned on time. The HVAC units were delivered on time. The subcontractor installed the equipment after the GC unloaded the units from the truck and hoisted them to location with their own crane. The system was completed and on a very hot (design day) day during balancing it is discovered that the units are not keeping up with air conditioning requirements. The labels are checked and it is discovered that they are undersized by one third from the specified units. Who is at fault? How could "active" quality control prevent this from happening? How is it resolved contractually and physically? How could the general contractor mitigate these costs?

Case 88:   WRONG CARPET

This is a small office tenant improvement (TI) project with a contract value of approximately $500,000. The owner is very experienced and routinely has one to two projects under construction. The architect is a repeat firm. The owner usually likes to negotiate projects but has recently had two bad experiences with general contractors and decides to bid this project out in hopes of finding a new firm they can develop a relationship with. The general contractor who is selected is looking for just this sort of arrangement. They are a very small firm. Even this project is at their limit. They operate very informally with little documentation. The project is on schedule and all parties are operating in a negotiated fashion. Samples of the carpet are not submitted as the specifications did not require this process. The carpet arrives two weeks before it is scheduled for installation. It is left rolled up and wrapped. The superintendent does not inspect it. The carpet was installed by the subcontractor over a weekend. The superintendent was not present during this work.

On Monday morning the owner arrives and is surprised to see that the new carpet is a different color and a different cut than the carpet it adjoins with and was to match. It turns out although that the carpet is "as specified" but the architect erred with their specifications. A new replacement carpet is fourteen weeks out. The furniture is coming off the truck and is ready for installation. The owner needs to move new people in this week.

What should the team do? Who pays for any rework? Did the client find their new general contractor team member? Should contractors submit only what is asked? Why would the design team not request that 'everything' be submitted? What happens if a contractor submits on a product for which a submittal was not requested? Will the design team review it? Will the architect's errors and omissions (E & O) insurance pay for removal and replacement and impact costs if this carpet is returned? Does the subcontractor have any liability?

_____
_____
_____
_____

SECTION 12: CHANGE ORDER PROPOSALS

Including Contract Change Orders

Case	Title
89.	Demobilized Demolition Subcontractor
90.	Subcontractor Change Orders
91.	Time and Material Change Orders
92.	New Chief Executive Officer
93.	Dirt Change Orders

Most of these case studies overlap with other primary topics and at least 21 other cases involve change order proposals. See Appendix 2 for a matrix connecting all 101 of the cases with all fourteen primary topics.

Case 89: DEMOBILIZED DEMOLITION SUBCONTRACTOR

Your position is the project manager for a general construction firm on a $100 million college laboratory remodel and expansion project. Your demolition subcontractor had a competitive bid of $1 million, just slightly lower than the second bidder. They mobilized quickly and performed the bulk of the original contracted demolition work ahead of schedule. Because of differing site conditions, the architect has requested pricing on numerous change orders affecting the demolition subcontractor's work. The merit of these changes is not in question, but the quantum is. The subcontractor sees this as a "contracting opportunity" and has inflated their pricing on the changes, totaling an additional $1 million. The owner employed a third party estimating firm to prepare check estimates. The owner offered the subcontractor $500,000 for the changes which were yet to be performed. The subcontractor did not like the offer and demobilized from the project, having completed all of the base contract work. There appears to be little you can do to motivate them to return. The client has indicated that this is your problem as they made a fair offer for the extra work. What can you do now to remobilize this subcontractor? Should you bring on another firm to perform this work? How can you do this contractually? Can you assure that the pricing will be fair? Should you accept the $500k change order and hire another subcontractor on a T&M basis? Can the owner force you to perform extra work at a value less than what you have requested?

_____
_____
_____
_____

Case 90: SUBCONTRACTOR CHANGE ORDERS

Two subcontractors approached change orders with a medical client differently. This was a complex lump sum project. The documents had errors and the owner also requested scope changes. The mechanical subcontractor does not charge the owner for small discrepancy changes but successfully collected on the substantial scope changes. The electrical subcontractor charged the owner for every $100 discrepancy. The owner took an immediate dislike to the electrical subcontractor; they burnt their bridge early in the project. The owner was tough on them even on the clear scope changes. The electrical subcontractor eventually claimed the owner at the end of the project. The mechanical subcontractor made a very fair profit. Was this a good approach from the mechanical subcontractor? Does it always work? Did the electrical subcontractor fail? If so, how? Where do you draw the line between give and take?

Case 91: TIME AND MATERIAL CHANGE ORDERS

A sole-proprietor developer has contracted with a general contractor to build a $20 million condominium complex. The developer has never constructed a project before and does not have relationships with either the contractor or the architect. The architect's services were terminated after receipt of the building permit. The developer acts as his own owner's representative. The contractor was employed through the competitive bid procurement method. There are numerous document discrepancies and differing site conditions. The contractor provides notice to the owner and proceeds with the most likely corrections. The owner is always pushing the contractor to speed the project up under advice from his Realtor that the market is hot and he needs to get the condominiums on the market. The changes are not estimated before the corrections are made but the contractor tracks their costs on a time and material (T & M) basis. The contractor does provide the owner with written notices that they are proceeding with each extra work item and will provide actual costs when they are completely realized.

Upon completion of the project, the condominium market has dropped off. The developer is unable to sell any of the units and ends up unloading the entire project to a large real estate trust at below cost. The contractor has submitted a claim for $500,000 for extra work as described above. All of the extras are legitimate and the pricing is substantiated with backup. The GC has not inflated or falsified any of its costs. Due to the expedited schedule, the project was difficult for the GC. If they collect this claim they will just break even on their costs. The first developer denies the extras under the basis that a) he did not receive proper notification, and b) he has lost money on the deal as it is and there is not anything left to share with the contractor. All parties in this case made numerous errors. What were they and what should they have done to correct this situation before it happened. Does the developer have any claim against his Realtor or the architect? Does the developer have a claim against the contractor? Can the GC recover from, or lien, the new owner?

Case 92: NEW CHIEF EXECUTIVE OFFICER

This is a negotiated team-build assisted living project. It is a complex remodel and addition. The patients are moved from wing to wing as the building is phased over a two-year construction period. The client, architect, general contractor, and major subcontractors all have experience working together. Half way through the project, the client's Chief Executive Officer (CEO) quits and a new individual is appointed by the Board of Directors. CEO number two has lump-sum construction experience. Several of his projects have resulted in claims and litigation. He distrusts all parties involved in the construction process, yet he enjoys the battles. He immediately begins to tell the Board how the previous CEO was not managing the project effectively and that all of the parties are taking advantage of the client. The project is on schedule, quality is adequate, there have not been any safety incidents, but there are numerous change orders outstanding. The CEO convinces the Board that he should hire a construction consultant to evaluate the change orders. How should the general contractor project manager deal with this situation? Should he team up with the other members of the design and construction team against the owner? If the project manager opens his books and works with the estimating consultant would this help his position? Should the PM make a run on the Board? Should he try to win over the new CEO and if so, what approach would you suggest?

Case 93: DIRT CHANGE ORDERS

93.1 A general contractor was the successful bidder on a $15 million lump sum three building office complex and has placed a very young and inexperienced team of project manager and superintendent out on the project site. The client is a sole-proprietor developer and a very experienced construction professional. The developer advises the contractor's officer-in-charge (OIC) of his concern of this young team. The OIC responds that this is a lump sum project and assignment of personnel is the GC's choice. He also assures the client that he will watch over his team. The project eventually goes astray for several reasons. The contractor loses money on the job and ultimately ends up dismissing both of the on-site staff. There are a significant amount of change orders, both agreed and disputed, and the project finishes three months behind schedule. Can a client in this situation request certain team members or require a change of personnel? Can the client now claim the contractor on the basis of an "I told you so?" What would you have done?

93.2 This same site is against a hillside, has a reputed active salmon stream through it, and it has a high water table. In addition, the project ends up with most of the earthwork performed in a record wet and rainy winter. There were three earthwork bids for this project. One subcontractor bids the earthwork at $1 million. The other two are close together at about $300,000. The high bidder is dismissed by the GC's young PM as obviously not needing the work and bidder number two is contracted. It turns out that there were only three borings described in the soils report. The Geotechnical firm had advised the developer that it would be better to perform at least ten borings, but due to cost reasons, the developer authorized only three samples. The report indicated that there was "some" water, and the dirt "may not" be suitable for back-fill unless kept "dry", and the contractor should anticipate "some" organic materials, but there was not any mention of debris. It turns out that not only is the site very wet, but it is completely a fill site. The earthwork is unsuitable for back-fill. The contractors discover substantial debris such as stumps, barrels, garbage, and even a Volkswagen Bug car body. None of the debris is contaminated. Have you ever read a soils report? Is the language specific or vague? What strategies do the geotechnical engineer, client, and even a bidding GC or earthwork subcontractor take with respect to soils reports?

93.3   It turns out that the high bidding earthwork subcontractor had dumped most of the debris at this site for a prior owner. It was not the subcontractor's workload that caused their high original bid, but rather insider knowledge. What could the general contractor have done to have better anticipated this situation? Is the young project manager negligent because he did not interview the high bidder? Is either the prior owner or the prior earthwork contractor liable? Is the Geotechnical engineer liable in any fashion?

93.4   The change orders for the earthwork noted above are significant and will eventually amount to $3 million. This is 20% of the original total bid amount from the GC. This of course quickly eats up any of the contingency the developer may have had. Why is earthwork so difficult to estimate? Are earthwork contractors as adept at estimating as are other subcontractors such as glazing or drywall? Do clients carry separate pools of contingency funds, such as some for earthwork and some for mechanical?

93.5 The inexperienced construction team decides that the best way to keep the original base earthwork contract and the change order earthwork separate is to hire a second earthwork contractor to perform the change order work. Is this customary? The original contracted earthwork firm essentially finishes their work, demobilizes, and is paid 100% of their contract. Later, when the owner claims that some of the change orders should have been assumed by the original subcontractor, there is nothing left of their contract to back charge against. Does either the client or the general contractor now have any recourse against this first subcontractor? The second firm performs the changes utilizing unit prices and time and materials cost accounting. Is this customary on a fast track project? How should change conditions be performed under a lump sum agreement?

93.6 The costs and the scope are substantiated for the second earthwork firm, but are they for the general contractor? Does the developer agree to these extra costs? If the developer doesn't agree, can the general contractor refuse payment to the second subcontractor? If this had been a Guaranteed Maximum Price (GMP) scenario, would the general contractor be allowed to pay for extra work from other sources of savings, even if they had something to do with the cost over-runs (which was not the case here)?

SECTION 13: CLAIMS

Including Dispute Resolution Techniques

Case	Title
94.	Nice Project Manager?
95.	Written Notifications
96.	Plumbing Claim
97.	School Electrician

Most of these case studies overlap with other primary topics and at least 16 other cases involve claims and dispute resolution. See Appendix 2 for a matrix connecting all 101 of the cases with all fourteen primary topics.

Who Done It: 101 Case Studies in Construction Management

Case 94: NICE PROJECT MANAGER?

This $80 million lump sum public high school was being built for a school district, which had never completed a project of this size. The elementary schools and remodel projects undertaken prior all were successful projects. They had never been claimed. The project manager for the general contractor was very experienced.

The owner consistently missed deliverable dates (permits, equipment, decisions on finishes) all of which were on the contractor's critical path. When this happened there was never an 'issue', rather the GC's superintendent would simply record it in red on the contract schedule hanging in the jobsite trailer conference room. In addition, when the architect returned submittals late, or lost RFIs, or missed meetings, the PM was very understanding and simply documented it in his meeting notes.

When the PM brought on another assistant PM to the job, along with a separate trailer full of empty file cabinets, the client also thought he was being very thorough. There had not been any emotional issues or nasty letters issued through most of the project. The documentation from the contractor's side was extensive. Was this really "the nicest project manager" the school district had ever seen? What do you think was going to happen with this project?

Case 95: WRITTEN NOTIFICATIONS

During the course of construction of a medical office building (MOB) the general contractor's project manager provided numerous verbal directions to the mechanical subcontractor requiring revisions to the schedule and changes to the plans and specifications. Many of these were required to resolve discrepancies and keep the project on schedule, which resulted in schedule compression. The general contractor and the subcontractor disagreed on pricing for some of these extra work items which were all subsequently negotiated into the subcontract. The subcontract contained the following clauses, which were pertinent to this case:

a. Subcontractor is not to perform work for which they have not received written direction

b. Subcontractor cannot submit a claim for extra work for which they had failed to provide timely and proper written notification

c. Subcontractor cannot submit a claim for damages associated with project delays caused by the owner, owner's agents, or the general contractor

95.1 Upon completion of the project the subcontractor submitted a claim, through their attorney, for $75,000 damages due to loss of productivity as a result of receiving multiple change orders and resultant schedule compression. The general contractor denied this claim based upon clauses 'b' and 'c' above. The mechanical subcontractor took the case to arbitration where the arbitrator found in favor of the subcontractor. The arbitrator ruled that the general contractor had, by their actions, waived the "in writing" clause 'b' and the "no damages for delay" clause 'c'. The arbitrator determined that the general contractor also invalidated clause 'a' when they verbally directed the subcontractor to perform extra work. In effect the arbitrator has indicated that because the general contractor did not follow the rules, the subcontractor was not required to follow the rules. He indicated the GC was "negligent and acted in bad faith." Is this fair to the subcontractor? Is this fair to the general contractor? What project management principles should both parties have used to prevent this from happening? Is arbitration the fairest way to resolve cases such as this?

___

95.2 Assume the role of the general contractor and prepare your case to appeal this ruling to an upper court. Does the contract allow you to appeal? Will this be expensive? Prepare an argument with at least five points utilizing whatever material you can gather, both from within and outside of the classroom and the contract.

Case 96: PLUMBING CLAIM

During the paper close out of your project, you receive a very well prepared, very professional claim from your plumbing subcontractor. This subcontractor worked very well with you throughout the project, although they were not the best with their paperwork. The quality of their work was acceptable. They received numerous change orders (worth over 20% of their original bid of $600k) and received approval of the majority of their requests for extra funds and extra time, after mutual negotiation. They signed your change orders. The plumbing project manager you have worked with throughout the project has been relocated and refuses to return your calls. You are now only hearing from their attorney. Assume a standard AIA GC-Subcontractor agreement. Their claim has the following planks:

a. Discrepant documentation resulting in extra costs: They received change orders for this work but usually at somewhat reduced value than what they requested. They request an additional $60k for this cause

b. The cumulative effect of multiple change orders: This has resulted in extra costs of $100k associated with loss of productivity and increase in jobsite general conditions

c. Schedule compression resulting in loss of productivity: They request an additional $90k for this issue

d. Extended schedule that caused additional field and home office costs of $110k

e. Their accounting records indicate that their actual costs ($1.08 million) over-ran their final contract value ($720k) by 50% ($360k)

96.1 Take the subcontractor's position: Why are you right? Use your project management tools and the contract.

96.2 Take the general contractor's position: Why are you right? Use your project management tools and the contract.

96.3 Be objective: How will this be resolved? What should you, as the GC's PM, have done to prevent this? What changes can you propose to the general contractor's change order and contract modification system to minimize these types of late, after the fact claims?

Case 97: SCHOOL ELECTRICIAN

An electrical subcontractor is filing three simultaneous claims against a public school district. The following facts are available

a. The GC's contract with the school district is $10 mil on a 9 month remodel project.

b. The electrical contractor's original contract value was $1.5 million.

c. The claims were filed at the completion of the project with only limited informal notice of any of the issues given prior.

d. The electrical subcontractor and the general contractor both bid the job lump sum.

e. The contracts that have been executed by all parties were standard AIA documents and were included in the specifications prior to the bid. They were executed without any language changes.

f. The general construction contract has $1000 per day liquidated damages. The general contractor contractually passed that risk on to all of its subcontractors.

g. The general and the subcontractors are all bonded.

h. The quality of the work and the safety performance were within standards and acceptable.

i. The project finished six weeks late, forcing the school into using temporary facilities well into October. The school district has realized real cost damages because of the delays totaling $500,000.

j. The reason behind electrician's <u>claim number one</u> is that the original contract schedule had a reasonable work plan that indicated the electrical subcontractor's planned manpower peaking for one month at two times the average manpower (8 VS 4) over the course of the project. The GC did not manage the project efficiently, used up the float shown in the schedule, and the electrical subcontractor ended up over-spending their labor estimate considerably and spent 90% of the actual manpower all during the last two months of the job. The theme of claim #1 is "inefficiency due to labor compression".

k. <u>Claim number two</u> is based upon the fact that the documents were at best 60% complete and were not coordinated between the different design disciplines. The client awarded change orders to the electrical subcontractor through the GC, which increased the subcontractor's original contract value by 50%. The electrician and the GC requested more funds throughout the change process but the client and the design team negotiated them down at every turn. The owner included language in the executed changes, which prohibited the GC from submitting additional claims for the added work. The GC included similar language to the subcontractor. The electrician is claiming "loss of productivity due to cumulative effect of change orders."

l. The <u>third claim</u> occurred half way through the project during electrical rough in. The GC discovered asbestos and the project was completely shut down for one month. All parties demobilized and re-mobilized one month later. A schedule extension was granted but the contract did not allow for extended general conditions. The subcontractor is claiming "loss of labor productivity" due to the fact that many of the key workers were lost during the month shut down, "extended field and home office general conditions", and "loss of fee potential".

m. The GC has processed all three claims through to the designer who has forwarded them on to the school. Neither the GC nor the designer took a strong position either way.

97.1   Take the position of the electrical subcontractor. Determine the value of all three claims. Use your estimate and schedule as backup. Present the claims using information you have learned from this manuscript, your classes, and outside professional experience. Perform outside research to reinforce your claims. You will only get one chance to present your position to the school board at a public hearing.

97.2   Take the position of the school board. Listen to the presentation of the electrical subcontractor. Reject all three claims. Use information you have learned in this manuscript and your classes as well as substantiating data you research from outside of the classroom. You will only get one chance to present your rejection. And remember, you want to be re-elected!

97.3   Because the school board and the electrical subcontractor could not agree (surprised?), the contract forced both parties into binding arbitration. Take the position of the arbitration panel. You are to decide in favor of one party or the other on each of the three claims separately. You do not need to find in favor of the same party on all three claims. You are to determine final award values. You are to base your decision on the presentation from each of the parties, information you have learned in your classes, information you research outside of class, and the documents. You are in the business of being hired as an arbitrator and cannot afford to have your positions questioned or overthrown by an upper court. They must be correct and sound.

97.3.1 Arbitration Ruling Claim 1:

97.3.2 Arbitration Ruling Claim 2:

97.3.3 Arbitration Ruling Claim 3:

97.4 Assume that the contract allowed for alternative dispute resolution (ADR) methods and specified that a dispute resolution board (DRB) be established before contract award. Explain what a DRB is and what benefits it provides over conventional dispute resolution methods.

97.5 Reviewing the presentation of the two parties and the contract documents, explain what "recommended" solution the DRB would make to the two parties. When should this DRB hearing be held? What could all four parties (including GC and designer) have done to keep these issues from occurring or, at a minimum, resolving the issues before formal claims were submitted post project? Could the presence of a DRB from the project's onset have prevented these issues from becoming an after-the-fact claim?

SECTION 14: ADVANCED TOPICS

Including Close-Out and Safety

Case	Title
98.	Close-Out Documents
99.	Who Closes?
100.	Highway Accident
101.	No Contractual Ties

Most of these case studies overlap with other primary topics and at least 9 other cases involve advanced topics such as close-out and safety. See Appendix 2 for a matrix connecting all 101 of the cases with all fourteen primary topics.

Case 98: CLOSE-OUT DOCUMENTS

A general contractor has successfully completed building a shipping-and-receiving facility. This project was initially procured as a competitive bid lump-sum project. This was a fairly simple $10 million, 100,000-square-foot tilt-up concrete facility. Total construction time was seven months. The owner, contractor, and architect have all worked well together throughout the project. All payment requests have been approved and paid on time. All change orders have been negotiated and incorporated into the contract. This owner apparently will have other similar projects coming up throughout the country and is interested in hiring the general contractor on a negotiated basis to provide all of their construction needs. The contractor has combined the last progress payment request for $100,000 with a request for retention release of $250,000 for a $350,000 final payment request. The general contractor moved their project manager off the project during the last month of the schedule and has turned over the close out of the project to a new project engineer fresh out of school. The project engineer has had a difficult time getting the close out documents together, especially the operation and maintenance manuals (O & Ms). The subcontractors are not responding with all of the required information. The architect has twice rejected the operation and maintenance manuals. Two months have now passed since completing the punch list work and receiving the certificate of occupancy from the city. The owner is now refusing payment of the $350,000. The architect has suggested that the general contractor separate out the $100,000 due and request that under a separate payment invoice. The project engineer is being stubborn and is refusing. What mistakes has the general contractor made? What can they do now to remedy the situation? Can the owner contractually hold the full $350,000 and if so, for how long? When will the general contractor's (and the subcontractors') lien rights expire? Is there any chance these parties can (or should) enter into a long-term national agreement?

Case 99: WHO CLOSES?

This typical school district was feeling the common pains of late projects and uncompleted punchlists, well into the school year. The outside party who serves as the owner's representative did not have an incentive to stay with a project after the certificate of occupancy was received. The general contractor was simply looking for release of the retention. The facility managers and maintenance crew did not have any construction involvement and did not have any contractual relations with any of the construction team. Most of the designer's fee is received prior to the permit stage and very little is left for the close-out process. Who should be responsible for proper close-out of a project? Do all of the team members have a built-in incentive for a speedy and sometimes poorly performed close-out phase? What about involving the school principal, don't they have the long-term incentive for a properly completed project? Should the end-user on any project be involved from design through construction and marshal a properly closed out project?

Case 100: HIGHWAY ACCIDENT

A flagger was killed on a highway construction project when a driver accidentally hits her. Unfortunately she was not paying attention and talking with a parked dumped truck driver when the accident occurred. The construction project appeared to be adequately signed and barricaded. The driver is also injured and his car is totaled as he swerved at the last minute to try to avoid the flagger. The woman was a young mother and her family's insurance company sues the driver, the contractor, and the state highway department (the client). The driver's insurance company in turn sues the deceased woman's family, the contractor, and the state highway department. Who can sue whom? We obviously know who lost in this scenario, but who wins financially? How can highway construction accidents be mitigated?

Case 101: NO CONTRACTUAL TIES

An out of town investment firm purchases a speculative office building complex while under construction. The buyer's contract is with the seller. The buyer does not have any contractual ties with either the contractor or the architect. The plans and specifications are part of the purchase-sale agreement. The escrow close occurs upon receipt of certificate of occupancy and completion of all punchlist work. The punchlist was developed by the seller. Two years after the close, there are numerous problems which show up associated with roof drains, storm drains, and the general slopes of site pavement and sidewalks. The contractor's one-year warrantee has expired. Upon investigation, the buyer is notified by the original civil engineer that the project was not built according to the documents in several instances. The engineer did not participate in development of the punchlist, as this was not in their contract. Does the buyer have a case against any of these parties?

ABOUT THE AUTHOR

Len Holm grew up in a construction family. His father Arne Holm was a high-end residential and light commercial general contractor in Grays Harbor County, Washington. Len was shoveling concrete and driving nails from the age of 10. He eventually became a journeyman carpenter and foreman. Len was the only member of his family to go to college and earned bachelor degrees in Building Construction and Economics and a master's degree in Construction Management all from the University of Washington. Len's first job out of college in the early 1980s was as an estimator and a scheduler traveling around the Country building power plants for one of the largest construction firms in the world. He later found his way back to Seattle and worked on numerous high-technology, medical, and industrial projects with a large local general contractor as a project manager, senior project manager, and company stockholder. Len founded his own company, Holm Construction Services, in 1994 which has provided a variety of construction consulting services on hundreds of residential and commercial projects including owner's representation, expert witness, and contractor training. He has been an instructor at the University of Washington since 1993 and has taught over 90 quarter-long courses on 13 different topics to over 2500 students. He has authored and co-authored several books and articles on a variety of construction issues including project management, estimating, and dispute resolution. Two books published by Pearson are standard textbooks for many construction management programs throughout the United States and in other countries: *Management of Construction Projects, A Constructor's Perspective*, and *Construction Cost Estimating, Process and Practices*, both co-authored with John Schaufelberger, PhD. Questions or comments regarding this manuscript may be sent to holmcon@aol.com. We hope you enjoyed the stories.

ABOUT THE ILLUSTRATOR

Barbara Holm..

APPENDIX 1: ABBREVIATIONS

The use of abbreviations is standard in the construction industry and many are used throughout this manuscript; most of them are listed here along with a few others.

ABC	Associated Builders and Contractors
ADR	Alternative Dispute Resolution, including DRBs and Mediation
AGC	Association of General Contractors
AIA	American Institute of Architects
BC	Back Charge
CCA	Construction Change Authorization
CCD	Construction Change Directive
CD	Construction Documents or Contract Documents
CEO	Chief Executive Officer
CM	Construction Manager or Construction Management
C of O	Certificate of Occupancy
CO	Change Order
C-O	Close-out
Contractor	General Contractor or sometimes Subcontractor
COP	Change Order Proposal
CP	Change Proposal
CPFF	Cost Plus Fixed Fee
CPPF	Cost Plus Percentage Fee
CSI	Construction Specifications Institute
D-B	Design-Build
DBE	Disadvantaged owned Business Enterprise
DDs	Design Development Documents
Demo	Demolition or Demolition Contractor
DeMob	Demobilization
DRB	Dispute Resolution Board
E & O	Errors and Omissions Insurance
FO	Field Order
FQ	Field Question (also Request for Information/RFI)
FWO	Field Work Order
GC	General Contractor, also General or Contractor
GCs	General Conditions or Multiple General Contractors
GC/CM	General Contractor/Construction Manager (Procurement and Contracting Method), Also CM/GC
General	General Contractor, also GC or Contractor

Appendix 1, Abbreviations, Continued:

GMC	Guaranteed Maximum Cost (Contract or Estimate), also GMP
GMP	Guaranteed Maximum Price (Contract or Estimate), also GMC
GWB	Gypsum Wall Board (Sub or Material), also Drywall or Sheetrock
HVAC	Heating Ventilation and Air Conditioning (Contractor or Subcontractor or Equipment)
JV	Joint Venture
K	1000, $40k = $40,000
LDs	Liquidated Damages
LR	Lien Release
LS	Lump Sum (Agreement, Bid, Estimate, Process)
M	Million (dollars), also Mil
MACC	Maximum Allowable Construction Cost (sim to GMP and GMC)
MBE	Minority owned Business Enterprise
M & E	Mechanical and Electrical (Subcontractors or Scope), also MEP
Mil	Million (dollars), also M
MOB	Medical Office Building
Mob	Mobilization
MXD	Mixed Use Development
NA	Not Applicable
NIC	Not In Contract or Not Included
O & Ms	Operation and Maintenance Manuals
OIC	Officer-In-Charge
PE	Project Engineer
PE	Pay Estimate, also PR – Pay Request
PL	Punch List
PM	Project Manager or Project Management
P.O.	Purchase Order (to a Material Supplier)
PR	Pay Request, also PE - Pay Estimate, or Public Relations
Pre-Con	Pre-Construction (Contract, Agreement, Process, Fee, Phase)
QC	Quality Control
QTO	Quantity Take Off
Rep	Owner's Representative
RFI	Request for Information (also Field Question)
RFP	Request for Proposal
RFQ	Request for Quotation or Request for Qualifications
ROM	Rough Order of Magnitude Estimate
SCHD	Schedule

Appendix 1, Abbreviations, Continued

SDs	Schematic Design Documents
SK	Sketch (numbered supplemental design drawing)
SBE	Small Business owned Enterprise
SOV	Schedule of Values (Pay Estimate Backup or Summary Estimate)
Spec	Specification or Speculation as in Real Estate Speculation
Sub(s)	Subcontractor(s) or Subcontract
Super	Superintendent
TI	Tenant Improvements
T & M	Time and Material (Cost Tracking or Contract)
TQM	Total Quality Management
U.P.	Unit Price (Estimate or Contract)
VBE	Veteran owned Business Enterprise
VE	Value Engineering
WBE	Woman (or Women) owned Business Enterprise
WBS	Work Breakdown Structure

APPENDIX 2: CASE STUDY MATRIX

This appendix lists all 101 of the cases in a table format and includes cross-references for many of the topics that each of the cases connects with. Many cases relate to other sections; most of the examples cross many topics. For example there are 5 cases, which have been categorized with Change Order Proposals (COP's) as the primary topic, but there are at least 26 which involve COP's to some extent. Scan down the left hand side of the matrix to find Section 12, Change Order Proposals. Now read across the top of the page to the COP column. Five cases are in Section 12 and are indicated with an 'A' which designates that COPs is their primary topic. A vertical scan will find 26 other cases with a 'B' which indicates that COPs are a secondary topic, or 'C' which indicates that these cases are also connected with Change Order Proposals.

Who Done It? 101 Case Studies in Construction Management
Appendix 2

APPENDIX B: CASE STUDY MATRIX

PRIMARY (A), SECONDARY (B), AND REFERENCED (C) TOPICS

Case No.	Case Title	Orgs. 1	Procure 2	Contract 3	Est. 4	Schd. 5	Subs 6	Startup 7	Comm's 8	Pay 9	Cost 10	QC 11	COP 12	Claim 13	Adv'd 14
	SECTION 1: ORGANIZATIONS, including project organizations, company organizations, and people issues														
1	Pass-Through Contractor	A	B	C			C								
2	Owner's Contracts	A			C		C								
3	CM or GC?	A		B				B							
4	Design-Build JV's	A		B										C	C
5	Owner's Subcontractors	A					B								
6	Bait & Switch	A		B				C							
7	Multiple Contracts	A					C								
8	Ambitious PM	A			B						C		C		
9	New Contractor	A	C												
10	Client Expertise	A													
11	Developer or GC?	A			B						C				
12	Overworked PE	A							C						
13	Self-Performed Failure	A		B	C		C			C				C	
	SECTION 2: PROCUREMENT														
14	Design-Build M & E		A	B	B		B			C				C	
15	Marketing	C	A												
16	Bid or Negotiate?		A												
17	GC/CM Accounting		A	B	B					C					
18	Bid @ 50%		A		B		B								
19	Union or Open Shop?		A												
20	Executive Home		A	C	B	B							B		
21	Successful Schools?		A	C			C				C		C		
22	Irrigation Union?		A	C			B								

Who Done It? 101 Case Studies in Construction Management
Appendix 2

Case No.	Case Title:	Section: Orgs. 1	Procure 2	Contract 3	Est. 4	Schd. 5	Subs 6	Start 7	Comm 8	Pay 9	Cost 10	QC 11	COP 12	Claim 13	Adv'd 14
SECTION 2: PROCUREMENT, continued															
23	Public Set-Asides		A												
24	Public Alternatives		A				C		C				C		
25	Labor Agreement		A											B	
26	Negotiated Success		A		C	C	A	B	C		B	C	C		C
SECTION 3: CONTRACTS, including insurance and bond issues															
27	Moving Target		B	A										C	
28	Budget or Bid?		C	A	B					C			B	B	
29	Turn-Key Impasse			A						C			B	B	
30	Historic Restoration			A	B	C	C		B	B	B		B	B	
31	Residential Dispute			A	B			B					B	B	
32	Shared Savings			A	B										
33	Subcontract Bonds			A	C		B						B		
34	All Inclusive?		C	A	C					B			B		
35	Seismic Repairs			A		C					C				
36	Line Item Estimate			A	B	B									
37	Allowance Accounting			A	B						C				
38	Contract Concerns		C	A	C										
SECTION 4: ESTIMATES															
39	No Sub Coverage		B	C	A		B								
40	Sub Short List		B		A		B								
41	Unknown Electrician				A		B								
42	Plan Centers				A		C								
43	Estimating Too Smart				A										
44	Window Bids				A		B								

131

Who Done It? 101 Case Studies in Construction Management
Appendix 2

Section:	Orgs.	Procure	Contract	Est.	Schd.	Subs	Startup	Comms	Pay	Cost	QC	COP	Claim	Adv'd
Case No. Case Title:	1	2	3	4	5	6	7	8	9	10	11	12	13	14

SECTION 5: SCHEDULES

45 Glazing Schedule					A	B						C		
46 Drywall Subcontractor					A	B					C			
47 Liquidated Damages					A	B							C	
48 Schedule Hold					A	B						C		

SECTION 6: SUBCONTRACTORS, including subcontracts, purchase orders, and suppliers

49 HVAC Union						A								
50 Union GC's						A								
51 Hospital Buyout	C			B		A								
52 Young Engineer	B					A		C						
53 Carpet Bankruptcy			C			A		C	B			C		
54 Steel Supplier					B	A		C	B				B	
55 First Team		C		C		A								
56 Concrete Walls			C	B		A	C					C		
57 Pulled Quote						A							C	
58 Hostile PM	C					A		C				C		
59 Fire Protection Heads						A		B			B			C

SECTION 7: START-UP, including pre-construction, mobilization, and value engineering

60 No Pre-Con Agreement			B				A							
61 20% Value Engineering				C			A							
62 Missed Pre-Con Meeting					C	B	A							
63 Early Mobilization			B	B			A							
64 Formal VE							A							
65 Sub Value Engineering						B	A							
66 Private VE			B			C	A							

Who Done It? 101 Case Studies in Construction Management
Appendix 2

Case No.	Case Title	Orgs. 1	Procure 2	Contract 3	Est. 4	Schd. 5	Subs 6	Startup 7	Comms 8	Pay 9	Cost 10	QC 11	COP 12	Claim 13	Adv'd 14
SECTION 11: QC, continued															
87	HVAC Units						B		B			A			
88	Wrong Carpet			B			C		B			A		B	
SECTION 12: CHANGE ORDER PROPOSALS (COP's), including contract change orders															
89	Demobilized Demo Sub						B						A		
90	Sub Change Orders						B						A		
91	T & M Change Orders			C					B				A	B	
92	New CEO								C				A		
93	Dirt Change Orders				B		B						A	C	
SECTION 13: CLAIMS, including dispute resolution techniques															
94	Nice Project Manager?													A	
95	Written Notifications					B	B						B	A	
96	Plumbing Claim					B	B							A	
97	School Electrician					B	B						B	A	
SECTION 14: ADVANCED TOPICS															
98	Close-Out Documents														A
99	Who Closes?									B					A
100	Highway Accident			B											A
101	No Contractual Ties			B											A

Note: Many cases have multiple sub-sections, parts, or positions, such as owner vs. GC or union vs. open shop